Erbers, Joan S.
Changing wilderness values,
1930-1990

980502

CHANGING
WILDERNESS VALUES,
1930–1990

Recent Titles in
Bibliographies and Indexes in American History

The People's Voice: An Annotated Bibliography of American Presidential
Campaign Newspapers, 1828-1984
William Miles, compiler

The Immigrant Labor Press in North America, 1840s-1970s: An Annotated
Bibliography
Volume 2: Migrants from Eastern and Southeastern Europe
Dirk Hoerder, editor
Christiane Harzig, assistant editor

The Immigrant Labor Press in North America, 1840s-1970s: An Annotated
Bibliography
Volume 3: Migrants from Southern and Western Europe
Dirk Hoerder, editor
Christiane Harzig, assistant editor

New Day/New Deal: A Bibliography of the Great American Depression,
1929-1941
David E. Kyvig and Mary-Ann Blasio, compilers

A Retrospective Bibliography of American Demographic History from Colonial
Times to 1983
David R. Gerhan and Robert V. Wells, compilers

Asian American Studies: An Annotated Bibliography and Research Guide
Hyung-chan Kim, editor

The Urban South: A Bibliography
Catherine L. Brown, compiler

Cities and Towns in American History: A Bibliography of Doctoral Dissertations
Arthur P. Young, compiler

Judaica Americana: A Bibliography of Publications to 1900
Robert Singerman, compiler

Shield of Republic/Sword of Empire: A Bibliography of United States Military
Affairs, 1783-1846
John C. Fredriksen, compiler

Biographical Index to American Science: The Seventeenth Century to 1920
Clark A. Elliott, compiler

The Natural Sciences and American Scientists in the Revolutionary Era:
A Bibliography
Katalin Harkányi, compiler

CHANGING WILDERNESS VALUES, 1930–1990

An Annotated Bibliography

COMPILED BY
JOAN S. ELBERS

Bibliographies and Indexes in American History,
Number 18

Greenwood Press
New York • Westport, Connecticut • London

Library of Congress Cataloging-in-Publication Data

Elbers, Joan S.
    Changing wilderness values, 1930-1990 : an annotated bibliography
/ compiled by Joan S. Elbers.
        p.   cm. – (Bibliographies and indexes in American history,
ISSN 0742-6828 ; no. 18)
    Includes indexes.
    ISBN 0-313-27377-4 (alk. paper)
    1. Human ecology – Moral and ethical aspects – Bibliography.
2. Nature conservation – History – Bibliography.   3. Nature
conservation—Moral and ethical aspects—Bibliography.
4. Wilderness areas – Bibliography.   I. Title.   II. Series.
Z5861.E4   1991
016.179′1 – dc20   91-7718

British Library Cataloguing in Publication Data is available.

Library of Congress Catalog Card Number: 91-7718
ISBN: 0-313-27377-4
ISSN: 0742-6828

First published in 1991

Greenwood Press, 88 Post Road West, Westport, CT 06881
An imprint of Greenwood Publishing Group, Inc.

Printed in the United States of America

The paper used in this book complies with the
Permanent Paper Standard issued by the National
Information Standards Organization (Z39.48-1984).

10 9 8 7 6 5 4 3 2 1

To the memory of my father, Erling
Swenson, who taught me to love
the wild things

CONTENTS

Preface   *ix*
Introduction   *xi*

BIOGRAPHY   1

HISTORY   9

PERSONAL, POLITICAL AND SCIENTIFIC STATEMENT   35

    Biology and Ecology   35
    Critiques of Western Culture   42
    Education   46
    Sierra Club Wilderness Conferences   49
    Wilderness and Wildlife Management   51
    Wilderness Defenders   55

PHILOSOPHY   65

POETRY AND FICTION   85

RESEARCH STUDIES   89

THE WILDERNESS EXPERIENCE   97

Author Index   *129*
Topic Index   *135*

PREFACE

I first conceived the idea for an annotated bibliography on wilderness values while reading in the magazine of the Wilderness Society that the twenty-fifth anniversary of the Wilderness Act of 1964 would be celebrated in September 1989. Since then I have spent many hours searching, reading and summarizing literature that was, initially, unfamiliar to me. This continuing effort has proven to be a cumulative learning process. I have discovered in the words and ideas of the writers included in this bibliography form and shape for my own strong, but inarticulate, feelings about the worth of the natural world. I have also come to better understand what seems to me to be the central problem of modern civilization: how to live with restraint and respect for other life and natural ecosystems in a time when human numbers and technological power have made our species a geological force.

My intellectual adventure could not have taken place without the help of many people. Montgomery College of Montgomery County, Maryland, supported my work by granting a semester's sabbatical leave for research. Bonita Favin of Rockville Campus Library of Montgomery College was invaluable in procuring books and articles for me. Even on the busiest of days, her commitment never faltered. I am grateful to Nana L. Anderson of the University of Utah Press and to Jean Monroe of Yale University Press for furnishing proof copies of books in the pre-publication process. My husband, Gerald Elbers, consistently encouraged my labors, even when they interfered with our daily life. I am grateful for his patience and for his dedicated proofreading.

INTRODUCTION

In September 1964 President Lyndon Johnson signed the Wilderness Act. For the first time in our history the American people committed themselves, in principle, to limiting the urbanization of the earth. Land was set aside, not as a game or hunting preserve or for spectacular scenery, but to be left alone. The new law mandated places where, in the words of the Act, "the earth and its community of life are untrammeled by man, where man himself is a visitor who does not remain." The Wilderness Act was the culmination of a long preparatory process, politically, involving years of lobbying by conservation organizations, and culturally, reflecting several decades of change in American thought about wilderness. It was also the beginning of a new intellectual endeavor, the philosophical re-evaluation of the relationship of humankind to nonhuman life and the land.

The ideas, political conflicts and sentiments that surround the Wilderness Act are recorded and reflected in a rich and varied literature. Aldo Leopold uses the phrase "round river" (originally the name of a legendary Wisconsin river in the Paul Bunyan stories) as a metaphor for ecological interconnectedness. Energy in an ecosystem flows "around and around in a never-ending circuit of life" (175). The literature of wilderness values, too, is a kind of round river, where the energy of love of the wilderness flows through the different types of wilderness writing connecting the passionate and the imaginative with the scholarly, scientific and philosophical. This bibliography is compiled to further that connection. There is, as yet, little evidence that contemporary policymakers are aware of the sophisticated work that is being done in environmental ethics; and even wilderness advocates seem underacquainted with the bulk of philosophical and historical writings on their subject. Certainly, there is no existing published work that reviews the literature and research on the attitudes toward wilderness which mold and transform our culture's relationship to the natural world.

The bibliography encompasses the different values Americans have sought in or attributed to the wilderness from 1930 to the present. It excludes, unless covered in collections or general works, some major earlier figures, such as John Muir and Henry David Thoreau and some major earlier movements, such as the resource conservation movement championed by Gifford Pinchot and President Theodore Roosevelt during the Progressive Era. The definition of wilderness employed for this bibliography modifies the definition given in the Wilderness Act to take account of scientific developments and environmental concerns that have surfaced since 1964. In addition to including the criteria of the absence of permanent dwellings and other artifacts of technological civilization and of the opportunity for solitude emphasized in the Act, this definition stresses what is only implied in the Act: wilderness is characterized by the presence of wild animals and plants carrying out their ecological roles within an ecosystem that maintains itself or undergoes natural succession without human control. There can be no wilderness without wildlife; there is only scenery. The very word "wilderness" is derived from the combination of the adjective "wild" with the Old English word for animal, "dēor." Wilderness is the place of wild beasts (Nash 34-35). Wilderness need not be pristine by human standards. Land that has been logged or mined or grazed by domestic livestock is classified as wilderness. Moreover, we know today that no part of the planet can escape the effects of acid rain, pesticide poisoning and chemical pollution generated by human activity.

Certain types of materials have been excluded from the bibliography. Readers will not find newspaper articles or articles from popular or general interest magazines. Articles from conservation magazines are included when they exhibit the viewpoint of the founders of the Wilderness Society or other important conservationists or when the articles present content not available in book form. Pictorial works whose primary purpose is to present the wilderness photographically are omitted. Although testimony at Congressional hearings relevant to the Wilderness Act represents the opinions about wilderness of thousands of people all over the country, over a period of several years, the sheer number of hearings made covering them excessively redundant. Book-length studies of particular species of wildlife that serve as a vehicle for exploring wilderness values have been included. Also, many annotated entries provide information about different species of wildlife and stress the importance of their role in wilderness experience and wilderness ecology.

Within the general scope of subject, time period and types of material the bibliography attempts to achieve completeness, offering information on the most relevant works. Unlike lists of authors' works or works about authors, where the relevance of an item is usually objectively clear, the appropriateness of an item to a study of wilderness values is not always immediately apparent from the subject cataloguing or from the title. A book may be relevant, even important, and seldom or never employ the word "wilderness." Reading is required, and that entails interpretation and judgment.

The books, essays and articles listed group themselves naturally into several broad categories: (1) Biography, (2) History, (3) Personal, Political and Scientific Statement, (4) Philosophy, (5) Poetry and Fiction, (6) Research Studies, and (7) The Wilderness Experience.

Biography. This section is composed almost entirely of books, although periodical articles have been included when no book source was available or when an article appeared to add

new information or express a special point of view. Biographies are given only for figures of importance in developing or
expressing wilderness values. Their number is further limited
by including only those biographies that serve to clarify such
values for the reader. Some attention has also been given to
the quality of the work. It seemed unnecessary to list popular
or series formula biographies when more complete, scholarly or
evocative publications were available.

History. This section lists historical treatments of wilderness or wildlife preservation movements, attitudes or
legislation together with histories of the idea of wilderness--
the secondary sources that form the core of any study of
wilderness values as they have emerged and developed over time.
In addition to these general studies, the section includes
histories of particular conservation struggles and of attitudes
towards wilderness in specific geographical areas, such as the
West. It is a category that can be comprehensively covered by
listing only books. However, a few articles that represent a
special insight or point of view are included.

Personal, Political and Scientific Statement. This is the
most diverse of the categories and the only category with
subdivisions. It brings together a variety of literature from
brief declarations of principle by wilderness advocates to
collections of papers from the Sierra Club Wilderness Conferences to knowledgeable explanations of the value of biodiversity by ecologists and other scientists. Most of the writings
included here attempt to influence or persuade. In contrast to
the history section, these are primary rather than secondary
sources--the expressions of opinion and thought that affect
politics and the public perception of the values of wilderness,
not the historical or philosophical analysis of these values.
However, natural history writing, the literary evocation of
place, that is, perhaps, most effective in bringing about
change in the way Americans view nature is covered in a
separate section devoted to "The Wilderness Experience."

Philosophy. This section surveys the attempts to deal
with the values of wilderness and natural ecosystems that
became part of formal philosophical discourse in the early
1970s. Although conservationists have often expressed the
values of wilderness, they have seldom analyzed the metaphysical or ethical justifications or implications of their positions. However, in the wake of increasing concern over
environmental degradation and extinction of species that took
place in the 1960s and early 1970s, academic philosophers began
to question the intellectual foundations of the world picture
that allowed these things to happen and to ponder the adequacy
of conventional ethical systems that locate value solely in
human society to solve environmental problems. The entries
brought together here represent the work of the environmental
philosophers who have dealt most directly with wilderness
preservation rather than, say, air pollution or nuclear war.
Even so, the selection has been difficult. On a broad criterion of relevance almost any article to have appeared in the
journal *Environmental Ethics* since its inception in 1979 might
have been considered pertinent. To avoid amassing a long list
of diffuse materials, I routinely excluded certain groups of
articles: articles narrowly focused on the critical interpretation of another environmental philosopher; articles that are
essentially arguments against the possibility of an environmen-

tal ethic; and articles by major writers that were later published in collections. In the latter case, however, articles or essays have been separately annotated if they were judged to have special value for understanding the author's theories. Articles that were annotated before they were published in book form have been retained.

Poetry and Fiction. This section lists works by American authors who have explored the value of wilderness in their fiction or poetry. Wilderness is a major focus in these works, and they should not be ignored in any study of wilderness values. It is important to make a distinction between these relatively few books and the many historical novels set in the frontier wilderness that have been published during the period since 1930. Although an interesting study might be undertaken to determine how they represent wilderness, none of the latter are included here.

Research Studies. This section covers reports and reviews of research, both surveys and controlled studies, that attempt to determine public attitudes and preferences concerning wildlife, wilderness and natural surroundings. Major book-length studies and collections form the core of the section. Periodical articles and essays in collections are annotated when they shed light on a topic not covered elsewhere, when they have a special relevance for wilderness values or when they review or summarize other research. I have not attempted to list the many limited studies of wilderness users conducted by the Forest Service and the Park Service over the past few decades. Readers can access these through George H. Stankey and Richard Schreyer's chapter in the *Proceedings of the National Wildlife Research Conference.*

The Wilderness Experience. This section covers nature writing that conveys in descriptive and evocative language the qualities of wilderness areas and natural ecosystems and how they contribute to helping define the human place in the scheme of things. The writers listed here describe the wilderness scene or wilderness adventure, relate its effect on them, and meditate on its significance. A salient characteristic of the works in this section is that they represent a distinct literary genre. They may contain philosophy and science, but they are not primarily scientific or philosophical treatises. Instead, they attempt to translate into language the aesthetic, emotional and sprirital qualities of the human response to nature. Ideas from three authors listed in the bibliography further define these writings. Paul Gruchow interprets natural history as governed by moral questioning about the "necessity to find our place in the natural world and to live within its limits" (216). Thomas J. Lyon identifies the three dimensions of nature writing as "natural history information, personal response to nature, and philosophical interpretation of nature" (3). Nature writers refocus our attention outward from the human species to the natural world (19). David Rains Wallace sees the task of the modern nature writer as evoking "appreciative aesthetic responses to a scientific view of nature" (112). Purely technical scientific descriptions, such as those that appear in field guides, and outdoors writings in which the wilderness serves chiefly as setting for recreational or occupational activities fall outside the parameters of this category.

The method used to assemble such diversity of material encompassing a single subject was eclectic. The *Subject Guide to Books in Print* and the *Cumulative Book Index* were researched early in the process of collecting citations; but it soon became apparent that the subject heading "wilderness areas" would retrieve only a small subset of relevant material, while broader headings, such as "nature conservation" or "natural history" retrieved large amounts of material of marginal relevance. A later search using these same headings in the OCLC database of library catalogs accessed through EPIC only confirmed this state of affairs. From a group of 350 entries published between 1988 and 1990, I could identify less than three percent as probably relevant. Nevertheless, two of the retrieved items proved to be essential sources still in the pre-publication process. Searches on DIALOG allowed me to refine my strategy by including variations of additional terms, such as "value," "attitude," "moral," or "ethics." Even so, these searches yielded far more irrelevant technical documents concerned with economic value than genuine items. A search for citations or references to Aldo Leopold unearthed some sources that I might not otherwise have discovered. Mary Anglemyer and Eleanor Seagraves' bibliography, *The Natural Environment: An Annotated Bibliography on Attitudes and Values* (1986), suggested some useful sources; but, in spite of the seeming relevance of the title, I found its coverage too broad to be of significant help in carrying out my task.

By far the most productive sources of citations were the notes and bibliographies found in the books and articles annotated in the various sections of my work. Special mention should be made of the books by Craig Allin, Linda Graber, Roderick Nash, Thomas J. Lyon, and Holmes Rolston III. Without the help of their knowledge of the literature the task of compiling this bibliography would have been prohibitively time-consuming. Other useful sources were the book review sections of certain periodicals, particularly, *Environmental Ethics, Wilderness*, and *Defenders*. *Environmental Ethics* is published by the Center for Environmental Philosophy at the University of North Texas, *Wilderness* is the publication of the Wilderness Society, and *Defenders* the magazine of Defenders of Wildlife.

The annotations are an essential part of the bibliography. They represent the distillation of many hours of reading over a period of almost two years. No item is included that was not first examined and read. Two purposes are served by the annotations. First, they save the user a great deal of time by enabling him or her to make an informed decision about the relevance of a particular item to the purpose at hand. Second, for the researcher who is willing to read an entire section or, better yet, the whole bibliography, the annotations will give an overview of concerned people, considered values, and significant political and intellectual movements and controversies. In other words, they serve to review the literature. To the inquiring student they should suggest possible topics for dissertations and research papers. The content of the annotations is largely descriptive, although they do contain some evaluative elements. For each item I have sought to delimit the scope of the work, identify overall themes and summarize major points. Evaluation enters the process when I assess the significance of the work and how it fits into the literature. It may also appear in the nature of the adjectives used to describe an author's arguments and organization. In annotating collections, where summary is not always possible, I have tried to emphasize special features or point out relevant parts. For poetry, fiction and nature writing, where literary qualities contribute heavily to the effectiveness of the work, I have attempted to convey something of the writer's technique and style by frequent use of brief quotations.

The bibliography is indexed by author and by topic. The main purpose of both indexes is to enable the user to locate related items that are classified in different sections of the work. They also provide quick access to single entries by an author or, occasionally, about a unique subject. Each work by an author appears as an entry number in the Author Index under his or her name regardless of whether it is a complete book, an essay in a collection or an article in a periodical. The only exceptions are forewords, introductions, conference discussions, writings by authors outside the time period, and certain pieces with little relevance to wilderness values that were included in otherwise pertinent collections. The Topic Index does not separately index items in anthologies.

In devising the Topic Index, I made no effort to completely catalogue each item by subject. Instead, I selected topical terms to facilitate what seemed to me to be the most useful alternate ways of grouping the entries. Geographical regions and features provided one focus for topical indexing. Whatever else wilderness may be, it is fundamentally place. Indexing by place names allows the user to collect sources for study of wilderness values as they are reflected in the history and appreciation of a region, such as, the Quetico-Superior, a state, Utah or California, for example, or a legislated entity like Yosemite National Park. Since this is a bibliography about values, I have tried to index items so that the user can identify material that explains or professes specific types of values. Aesthetic, scientific and heritage values are among those indexed, as are the more technical ecocentric, biocentric and anthropocentric values. Ecocentric refers to values located in wholes, such as ecosystems, biocentric to values located in living beings, and anthropocentric to values located solely in human beings. Works that list and explain many kinds of values are indexed under the term "values classifications." When values are interwoven or poetically expressed, however, as they often are in the section entitled "The Wilderness Experience," they have frequently proven impossible to index.

A few other index terms require elucidation here. The term "Western world picture" is used to bring together those works that discuss the implications of traditional philosophic, religious and scientific ideas for ecological degradation and destruction of wilderness. "Cultural change" is used for reference to works that interpret the present as a time of radical change in the way humankind relates to the natural world. The meaning of other terms in the Topic Index should be clear to the user.

Something further needs to be said about the chronological scope of the bibliography. The year 1930 as a beginning date was not simply an arbitrary cut-off point, nor was it chosen at the time the bibliography was contemplated. Instead, it emerged as the result of reading in the literature. As my research into wilderness values wilderness deepened, it became apparent that change was accelerating towards the end of the

1920s. The resource management theory of conservation epito-
mized by Gifford Pinchot, Theodore Roosevelt's chief of the
Forest Service, had advocated conservation of physical resources, such as forests, for the purpose of supplying lumber for
future generations. It was characteristically utilitarian and
anthropocentric, emphasizing the greatest good for the greatest
number of human beings; and that good was overwhelmingly
perceived as material. Though still the official philosophy,
by 1930 it was beginning to be challenged by notions of
permanently preserving nature for recreation and spiritual
renewal. New organizations devoted to protecting wildlife and
wilderness were being organized and gaining adherents among the
general population. The Wilderness Society was founded in
1935, and the National Wildlife Federation in 1937. Ecology,
which would form the scientific girding both for Aldo Leopold's
concept of a land ethic and for the philosophical exploration
of the intrinsic value of wilderness that emerged in the 1970s,
became a full-fledged science when Arthur Tansley developed a
quantitative ecosystem model in 1935. Even earlier, in the
late 1920s, ecologist Charles Elton had developed the model of
biotic communities, in which each species occupied a trophic
niche, that inspired Leopold's thought. Most important in
February of that year Bob Marshall published his seminal
article, "The Problem of Wilderness." Here at the beginning of
our period appears this perceptive, outspoken leader's call to
arms, asking all friends of the wilderness ideal to unite in
pressing for legally protected wilderness areas. Sixty years
later and more than twenty-five years after the Wilderness Act
it remains a call that inspires defenders of wilderness.

Works Cited

Gruchow, Paul. *The Necessity of Empty Places.* New York: St. Martin's, 1988.

Leopold, Aldo. "The Round River." *A Sand County Almanac, Other Essays from <u>Round River</u>.* New York: Oxford University Press, 1966

Lyon, Thomas J. *This Incomperable Lande: A Book of American Nature Writing.* Boston, MA: Houghton Mifflin, 1989.

Nash, Roderick. "'Wild-Dēor-ness,' the Place of Wild Beasts." *Wilderness: The Edge of Knowledge.* Proceedings of the 11th Wilderness Conference. San Francisco, 1969. Ed. Maxine E. McCloskey. San Francisco, CA: Sierra Club, 1970

Wallace, David Rains. "The Nature of Nature Writing." *The Untamed Garden and Other Personal Essays.* Columbus, OH: Ohio State University Press, 1986.

CHANGING WILDERNESS VALUES, 1930–1990

BIOGRAPHY

1 Brooks, Paul. *The House of Life: Rachel Carson at Work.*
   Boston, MA: Houghton Mifflin, 1972.
A moving chronicle of Rachel Carson's literary and professional life written by her editor and friend. The author
uses the literary device of interspersing excerpts from
Rachel Carson's own writings among his accounts of the
development of her books to reinforce the reader's
appreciation of her conscientious scientific accuracy and
her inspired craftsmanship. Brooks is especially successful in depicting the determination to reveal the basic
irresponsibility of technological society towards the
natural world that enabled her to endure the years of
exhaustive research, family tragedy, and ill-health that
preceded the publication of *Silent Spring.* She put aside
her plans for a major work on the evolution of life and
turned away from the beauties and mystery of the wild in
order to write this famous book for which she was cruelly
attacked and mocked by the chemical manufacturers and the
agribusinesses and their scientific representatives.

2 Brower, David. *For Earth's Sake: The Life and Times of*
   David Brower. Salt Lake City, UT: Gibbs Smith,
   1990.
The first volume in a projected complete autobiography of
one of the most active and influential leaders of the
conservation movement in the decades since World War II.
David Brower's book is not so much a chronological narrative as a collage of present reminiscences and past
writings organized in broad topical areas, for example,
family, friends, mountaineering, etc. The earliest
writing included is an account of a 48 day wilderness
climbing trip in the Sierras in 1934, which originally
appeared in the *Sierra Club Bulletin*; the latest is his
testimony on the Arctic National Wildlife Range printed
in the *Congressional Record* for October 15, 1987. The
section entitled "Why Wilderness?" reprints the author's
forewords and remarks from the proceedings of the Biennial
Wilderness Conferences sponsored by the Sierra Club from
the late 1950s through the early 1970s. Brower's clear,

closely argued expressions of wilderness values remain
remarkably contemporary even today.

3   Callicott, J. Baird, ed. *Companion to a Sand County
        Almanac: Interpretive and Critical Essays.* Madison,
        WI: University of Wisconsin Press, 1987.
    A collection of essays intended to "enrich and extend a
    reading of Leopold's classic." The pieces are organized
    according to their areas of emphasis. The first section
    is biographical, while later sections focus on *A Sand
    County Almanac* as a literary text and on its philosophical
    ideas or its impact on environmental ethics and conserva-
    tion practices. The collection also reprints Leopold's
    original 1947 introduction to the book he then called
    "Great Possessions." The essays that deal most directly
    with the values of wilderness have been annotated else-
    where in this bibliography, but students of the subject
    would also benefit from reading the literary critiques by
    John Tallmadge and Peter A. Fritzell.

4   Darling, F. Fraser. *Pelican in the Wilderness: A
        Naturalist's Odyssey in North America.* New York:
        Random House, 1956.
    The lively record of Scots ecologist F. Fraser Darling's
    study trips to the wilderness areas of the United States
    in the early 1950s. Darling's visits were sponsored by
    the Rockefeller Foundation and involved warm and continu-
    ing friendships with important figures in American
    conservation like Starker Leopold, Paul Sears and Olaus
    Murie, as well as the companionship and mutual respect of
    the lesser known wildlife experts who guided him through
    Wisconsin, Alaska and the West. At the Leopold farm he
    plants a commemorative burr oak and listens to young
    Estella play the guitar as they sit one evening around a
    campfire outside the famous Shack. He spends a week with
    the Muries in Wyoming examining the condition of the elk
    range in Yellowstone and taking long walks with Gurie, the
    Murie's coyote pup. Darling delights in America and in
    Americans; but his ecologist's eye cannot help but see the
    devastation wrought by human misuse and mismanagement of
    the land. Only a conservation ethic that will raise the
    level of human consciousness about our effects on the
    natural world can regenerate the earth.

5   Darlington, David. *In Condor Country.* Boston, MA:
        Houghton Mifflin, 1987.
    A biographical portrait of brothers Eben and Ian McMillan,
    two elderly ranchers and radical environmentalists of the
    San Luis Obispo country. David Darlington accompanies
    Eben on long treks through BLM and national forest land
    to view an abandoned condor nest and catch a glimpse of
    the rare California condor itself circling in the sky.
    He comes to understand the McMillans' conviction that
    saving the condor and other endangered species is not so
    much a matter of protecting individual animals through
    captive breeding as it is of respecting the ecosystem and
    accepting the necessity to work within it. The McMillans
    believe that the probable extinction of the condor is
    largely the result of the mentality that uses poisons to

control predators, rodents and insects. Condors ingest deadly 1080 and DDT (smuggled from Mexico) from the carcasses they feed on. Still, the brothers fight on for the condor and other wild things, inspiring younger people to follow in their footsteps. As Ian says, "Unless you fight, there'd be no chance of saving anything."

6   Flader, Susan L. *Thinking Like a Mountain: Aldo Leopold and the Evolution of an Ecological Attitude Toward Deer, Wolves, and Forests.* Columbia, MO: University of Missouri Press, 1974.
A detailed study of the contrasts in Aldo Leopold's thought between the early years, with their emphasis on game management for sustained yield, and the concern with the ecological integrity of the land he expressed in later years. The year 1935, the year that Leopold joined in founding the Wilderness Society, marked the reorientation of his thinking from conservation as balancing supply and demand to conservation as preserving and restoring the ecological health of the land. The author views this change as both contributing to and mirroring the history of evolutionary and ecological thought in the 20th century. Leopold's long puzzlement over the problems of deer population eruptions, habitat degradation, and elimination of predators both in the Southwest and in Wisconsin is extensively documented.

7   Fox, Stephen. "'We Want No Straddlers'." *Wilderness* 48.167 (Winter 1984): 4-19.
The history of the first fifty years of the Wilderness Society told through vignettes of the founders and other significant contributors to its growth and development. The title quotes Bob Marshall's criterion for the people who would be asked to join the nascent society in 1935. Fox uses it to describe the dedication to wilderness preservation that, however varied their backgrounds and personalities, characterized the leaders of the organization from Robert Marshall and Benton MacKaye to William Turnage, the executive director on the society's fiftieth anniversary.

8   Glover, James A. *Wilderness Original: The Life of Bob Marshall.* Seattle, WA: Mountaineers, 1986.
The first book-length biography of wilderness advocate and forester, Robert Marshall. The author shows Marshall's unprecedented dedication to organizing conservationists to band together to fight for wilderness preservation as the product of his father's political activism for minority rights and his childhood at Knollwood, the Marshall's summer place in the Adirondacks. It was also the product of his remarkably dynamic personality. Glover portrays Bob Marshall's humor, energy and enthusiasm, whether climbing mountains or entertaining friends, as the basis for the successful launching of the Wilderness Society. Marshall died in 1939 at the youthful age of 38; yet he remains today, perhaps, the most admired and loved hero of the wilderness preservation movement.

9   La Bastille, Anne. *Women and Wilderness*. San Francisco,
            CA: Sierra Club Books, 1980.
        A history of women's attitudes towards wilderness culmi-
        nating in profiles of fifteen modern women who have made
        their lives and careers out of contact with wild nature.
        The fifteen were chosen as representing the broadest scope
        of occupations, locations, ages, marital status, and
        wilderness philosophy from more than 350 names gathered
        from colleagues in the natural sciences, wilderness-
        oriented organizations, and responses to an ad in *Back-
        packer* magazine. Anne La Bastille prepares the ground for
        her biographical sketches by tracing the history of women
        on the western frontier. Most of these 19th century women
        entered the wilderness reluctantly and found it an alien
        land. Today's wilderness women have deliberately chosen
        their work and uncommon lifestyles. Whether biologists,
        sportswomen, writers, or educators, all share a deep
        appreciation of the natural world and an ability to be
        comfortable in the wilderness. Many are ardent and
        outspoken conservationists. A few seem to value wilder-
        ness chiefly as a training ground for self sufficiency and
        survival skills. La Bastille believes women are in the
        wilderness to stay and will be increasingly represented
        in outdoor and ecological professions.

10   Margolis, John D. *Joseph Wood Krutch: A Writer's Life*.
            Knoxville, TN: University of Tennessee Press, 1980.
        A beautifully written biographical study of the major
        theme in Krutch's life--the conversion from small town boy
        to urban intellectual to celebrant of the natural world.
        John Margolis shows Krutch breaking with the pattern of
        the fashionable intellectual when he refuses to replace
        the skeptical modernism of the 1920s with the utopian
        dogmas of the 1930s. Krutch develops his own personal
        stance out of his increasing interest in nature and his
        biographical work on Samuel Johnson and Henry David
        Thoreau. The happiness and productivity of his later
        years in Arizona reflect his success in creating an
        authentic life.

11   McPhee, John. *Encounters with the Archdruid*. New York:
            Farrar, Strauss and Giroux, 1971.
        A portrait of Dave Brower, a leader and spokesperson in
        the campaigns for wilderness preservation in the 1960s and
        an enthusiastic mountaineer. The author reveals Brower's
        dedication and purity of purpose but also his congeniality
        as a companion on wilderness outings by relating his
        conversations and interactions on three different trips
        with men who disagree with his stance but enjoy the
        outdoors: a mining geologist, a developer of resort
        communities, and a dam-building U. S. Commissioner of
        Reclamation.

12   Meine, Curt. *Aldo Leopold: His Life and Work*.
            Madison, WI: University of Wisconsin Press, 1988.
        A detailed and comprehensive biography of Aldo Leopold's
        disciplined and productive life. The author illuminates
        events and influences heretofore known only in their
        broadest outlines: his family background; his determined

courtship of Estella Bergere; his childhood and school life at Lawrenceville and Yale; the fortuitous meeting with Charles Elton, the author of *Animal Ecology* in 1931; the 1934 trip to Germany where he became acquainted with the disastrous effects of ecologically ignorant management of forest soils and wildlife. Meine's most valuable contribution to an understanding of Leopold's life and thought is cumulative--the thorough documentation that the change from conservation as preservation of hunting grounds to conservation as land ethic in Leopold's thinking was not an easy rhetorical passage but a hard won conversion based on accumulated research, personal experience and intellectual struggle. This study is the definitive biographical source on Leopold.

13   Mitchell, John G. "In Wilderness Was the Preservation of
          a   Smile: An   Evocation   of   Robert   Marshall."
          Wilderness 48 (Summer 1985): 10-21.
An appreciation of Robert Marshall, one of the founders, on the fiftieth anniversary of the Wilderness Society. The author pays tribute to Marshall's courage and sense of humor and to his enthusiastic, straightforward love of the wilderness.

14   Stegner, Wallace. *The Uneasy Chair: A Biography of Bernard
          De Voto.*   Garden City, NY: Doubleday, 1974.
The story of Bernard De Voto's life and times told by his friend and fellow westerner, conservationist and writer, Wallace Stegner. Although De Voto began his career as an alienated, insecure westerner seeking literary eminence in the East, he never lost his interest in western history or ceased to speculate about its meaning. Stegner sees De Voto's career as culminating in 1947, the year he published his award-winning history *Across the Wide Missouri* and the year he took on the western stockmen, the "Two-Gun Desmonds," as he called them, in the postwar controversy over western public lands that would "enlist his heart, conscience, and knowledge for the rest of his life." He used his position as a feature editor for *Harper's*, his wide contacts in the federal agencies, his reputation as a scholar and the strength of his prose to defeat a landgrab, which would have, in effect, conveyed the public lands of the West to a group of wealthy ranchers for a pittance. Old journalism usually makes dull reading; but, as Stegner points out, De Voto's work lives as an example of how to conduct the endless fight for preservation against opponents who are always the same.

15   Strong, Douglas H.   *Dreamers and Defenders: American
          Conservationists.*   Lincoln,   NB: University   of
          Nebraska Press, 1988
Readable summaries of the contributions of famous conservationists to the environmental movement. Strong's portraits emphasize the differences in values and interests among the people who are generally recognized as leading conservationists. He is especially informative in calling attention to the importance of the political philosophies of Harold Ickes and Barry Commoner in

determining their support for various environmental
causes. His coverage of Dave Brower sheds more light on
the controversies surrounding Brower's defeat for re-
election as executive director of the Sierra Club in 1969
than might be expected in a collective biography. The
chapters on Aldo Leopold and Rachel Carson, however, as
well as those on earlier conservationists, are useful
chiefly as introductions for readers seeking an initial
acquaintance with seminal figures in conservation history.

16   Tanner, Thomas, ed. *Aldo Leopold: The Man and His
        Legacy*. Ankeny, IA: Soil Conservation Society of
        America, 1988
A collection based on papers presented at the Aldo Leopold
Centennial Celebration held at Iowa State University in
October 1986. The contributions are grouped in three
major sections. The first includes essays by scholars
documenting the development of Leopold's land ethic and
its significance for American life and thought. The
contributors to this section are Susan Flader, Craig W.
Allin, Curt Meine, Roderick Nash and J. Baird Callicott--
all respected writers of conservation history and philosophy. In the second section people in conservation
careers write about the influence Leopold has had on their
lives and on the professions of   ecology and wildlife
management. The third section is composed of reminiscences of Leopold's children and his younger brother
Frederic. This anthology is a good general introduction
for understanding Leopold as a central figure in transforming wilderness values and for recognizing his contributions to preserving wildernesses on Forest Service
lands.

17   Terres, John K., ed. *Discovery: Great Moments in the
        Lives of Outstanding Naturalists*. New York:
        Lippincott, 1961.
An anthology of essays written in response to a letter
sent by editor John Terres to professional naturalists
requesting accounts of their memorable experiences
discovering facts about nature. Their replies relate
happenings in places throughout the world and involve
birds, mammals and reptiles, large and small, rare and
ordinary. Nevertheless, these field scientists share a
common fascination with nature and a strong awareness of
what zoologist Walter Taylor describes as the "great
stream of matter and energy which flows restlessly through
man, his living associates, and, in fact, through all
nature and the universe." Some stories have an almost
mystical resonance. After a night spent meditating on
Mount Sinai, Colonel Meinertzhagen turns to see a lammer-
geier, a bird he has long sought and studied, perched
"gloriously golden in the sunlight on the altar at the
summit. John Kiernan describes a moonlit night in
Flanders when, between the cacophony of German bombing
raids, he hears the exquisite song of a nightingale. F.
Fraser Darling recalls the intense childhood perceptions
and sense of wonder that led him to become a naturalist.
Most of the respondents evidence a close connection

between their commitment to science and their dedication
to conservation.

18   Vickery, Jim Dale. *Wilderness Visionaries*. Merrillville,
         IN: INC Books, 1986.
      A collection of biographies of important wilderness
      writers focused so as to reveal their contributions to the
      evolution of the meaning of wilderness to North Americans.
      No attempt is made to be complete; this is a very personal
      book.   Part of the author's preparation for writing it
      involved climbing the mountains and paddling the lakes and
      rivers his subjects loved and   wrote about.   He is most
      successful in interpreting the men and the country he
      knows best--the Quetico-Superior wilderness and Calvin
      Rutstrum and Sigurd Olson.   Until a book-length biography
      of Olson is published, Vickery's short study will remain
      an accessible summary of his life and thought.

19   Watkins, T. H.   *Righteous Pilgrim: The Life and Times of
         Harold L. Ickes*, 1874-1952.   New York: Holt, 1990.
      A thoroughly documented yet readable biography of the
      energetic and persistent man the author calls "one of only
      three Interior secretaries in American history to under-
      stand fully and value the importance of wilderness
      preservation to the spiritual and ecological well-being
      of the nation."   Harold Ickes did not become Franklin
      Roosevelt's Secretary of the Interior until he was 59
      years old; so much of this massive book deals with his
      struggle to rise from poverty, his commitment to Progres-
      sive reform and minority rights, and his dedicated
      immersion in campaign politics.   Wilderness becomes a
      major theme in Part VII, "A Department of Conservation,"
      where T. H. Watkins portrays Ickes as the heir to the then
      shattered and disorganized preservationist wing of the
      conservation movement.   Under his leadership at Interior
      the National Park System grew from "8.2 million acres to
      more than twenty million," including the newly founded
      Olympic, Great Smoky Mountains, and Everglades national
      parks.   Moreover, while adding land to the parks, he
      opposed adding roads.   In Ickes opinion the parks were for
      people who wanted to renew their communion with nature by
      camping and hiking.   He even supported legislation that
      would give the President power to set aside wilderness
      areas within the parks by proclamations that could be
      overturned only by acts of Congress.   In this he was more
      than a quarter century ahead of his time.

20   Watkins,   T.   H.   "Typewritten   on   Both   Sides: The
         Conservation Career of Wallace Stegner."   *Audubon*
         September 1987: 88-103.
      An overview of the significance of "one of the central
      figures in the modern conservation movement."   Watkins
      weaves a litany of Stegner's writings into the history of
      wilderness preservation since the 1940s to reveal Steg-
      ner's importance in articulating the postwar realization
      that progress is not an unrelieved good   and that the
      highest use for the land is often just to leave it alone.
      He   is   especially   persuasive   in   his   contention   that
      Stegner's 1960 letter to the Outdoor Recreation Resources

Review Commission became the manifesto for the movement that lead to the Wilderness Act.

HISTORY

21   "A Handbook on the Wilderness Act." *The Living
        Wilderness* 86 (Spring-Summer 1964).
      A special issue, which includes the text of the Wilderness
      Act of 1964 plus a legislative history and record of the
      Wilderness Bill as introduced in the 88th Congress. The
      issue contains statements by elected officials and
      environmentalists involved with the bill and features Dave
      Brower's tribute to Howard Zahniser of the Wilderness
      Society, one of the most effective proponents of the idea
      of legislatively designated wilderness areas, who died
      shortly before the act was passed.

22   "Wilderness: Past, Present, and Future." *Natural
        Resources Journal* 29.1 (Winter 1989).
      A special issue devoted to state-of-the-art papers that
      assess developments in wilderness practices and theory
      worldwide. George H. Stankey outlines the history of the
      concept of wilderness in Western culture through the 19th
      century; Robert E. Manning classifies the contemporary
      wilderness values that are implicit in the Wilderness Act;
      and Robert L. Lucas summarizes recent data on wilderness
      recreation in the national parks and national forests.
      Other contributors discuss wilderness preservation in
      Australia, Canada and throughout the world. Philip
      Dearden's provocative summary article, "Wilderness and Our
      Common Future," argues that, in the increasingly overpopu-
      lated world of the future, wilderness will be valued
      chiefly for its biophysical qualities, its ability to
      regulate essential life-processes, preserve wildlife and
      act as a genetic reservoir, rather than as a recreational
      setting or a place to experience solitude and closeness
      to nature. Moreover, wasteland wilderness areas will not
      be sufficient to protect the biosphere. Instead, nations
      must forego the immediate benefits of development and
      deliberatively set aside economically valuable land for
      the long range protection of the global environment. The
      affluent nations of the West will have to bear much of the
      cost associated with this change in societal values.

23   Ahlgren, Clifford, and Isabel Ahlgren. *Lob Trees in the Wilderness.* Minneapolis, MN: University of Minnesota Press, 1984.
An exploration of the relationship between changing attitudes towards wilderness and the effects of human activity on the Boundary Waters Canoe Area Wilderness. Each chapter concentrates on a particular human activity-- trapping, big pine logging, pulp wood logging, mining, recreation, etc.--as it is connected with one of the major tree species of the northwoods ecosystem. The authors see the BWCA as a land between the isolation and great expanses of western wildernesses and the smaller, easily accessible wilderness areas of the east. In spite of its pristine appearance to the casual eye, it is a land that has been altered by human actions and buffeted by differ- ing concepts of wilderness, from fear and distaste, through grab-and-get-out resource mining, to watershed conservation, recreation, and, most recently, primeval solitude and intrinsic value. The Ahlgrens are convincing in their conviction that maintaining the natural ecosys- tems of this area so changed by past human activity and subjected currently to such heavy recreational use will require scientifically informed and subtle forestry practices and management. They ask the question, "Can we learn to walk softly with a healing touch in this already injured ecosystem...?" Their book is an excellent source for enlightening the debate over the place of wildlife and forest management in the wilderness system.

24   Allen, Thomas B. *Guardian of the Wild: The Story of the National Wildlife Federation, 1936-1986.* Bloomington, IN: Indiana University Press, 1987.
The history of the National Wildlife Federation from its founding under Jay "Ding" Darling in 1936. Although this is a straightforward, chronological narrative, it does make clear the important influence of the NWF's educa- tional and marketing efforts in creating a broad constitu- ency for wildlife habitat protection and other environ- mental legislation. The appendices include a list of NWF presidents, a list of winners of various NWF conservation awards, and an annotated list of litigation in which the NWF participated.

25   Allin, Craig W. *The Politics of Wilderness Preservation.* Westport, CT: Greenwood, 1982.
A well researched history of the politics of wilderness preservation. The first two chapters cover the period through 1955. Later chapters cover the Wilderness Act of 1964, its implementation and its legacy. A special section examines the controversy over the Alaskan wilder- ness. This book is especially useful for study of changing wilderness values in that it recounts the arguments for preservation expressed by wilderness proponents at each political or legislative crisis. The author's overall development of the topic illustrates the thesis that, over time, the United States has moved from a period of natural resource abundance, labor-intensive technology and low population density, which placed little value on wilderness, to a period of relative resource

scarcity, high technology, and affluence, where wilderness is valued for its rarity.

26   Bean, Michael J.   *The Evolution of National Wildlife Law.*   Revised and expanded edition.   New York: Praeger, 1983.
An authoritative, highly readable history of federal wildlife law from its antecedents in the medieval doctrine that wilderness and wild animals belonged to the king through the 19th century interpretation of the government's right to regulate interstate commerce to the broad role for protecting wildlife embedded in recent statutes. Bean traces three major themes in federal wildlife law-- restrictions on taking, regulation of interstate commerce, and protection of habitat--as these have emerged in specific statutes, developed in judicial interpretations, and culminated in the Endangered Species Act and other comprehensive acts of the 1960s and the 1970s.   He also examines laws, for example, the Wilderness Act and the NEPA, that have important implications for wildlife preservation, although they do not specifically address wildlife issues. In analyzing the historical significance of the changes in wildlife law over time, he identifies two major trends: the expansion of the range of protected wildlife to include any member of the animal kingdom and the expansion of legally recognized wildlife values beyond that of food resource to include recreational, symbolic, educational, ecological and aesthetic values.

27   Brooks, Paul.   *How Literary Naturalists from Henry Thoreau to Rachel Carson Have Shaped America.*   Boston, MA: Houghton Mifflin, 1980.
A review of how nature writers have shaped our perceptions of the value of the natural world.   The author uses a broad historical organization to trace the changing role of the literature from early writers calling attention to the beauty and grandeur of wild nature to later writers identifying ecological relationships or exposing the destructiveness of technological culture.   Within this chronological framework individual writers are compared and contrasted in chapters devoted to single themes, the wilderness ideal, for example.   Brooks has a talent for succinctly summarizing the strengths and contributions of each writer he discusses.   His book is an enjoyable reminder of how much we owe to these writers who felt deeply enough about the integrity of nature to share their knowledge, their experience and their passion.

28   Brooks, Paul. *The Pursuit of Wilderness.*   Boston, MA: Houghton Mifflin, 1971.
A record of key battles for wilderness preservation that took place in the late 1960s.   Paul Brooks visited threatened wilderness sites in Florida, Washington and Alaska to ensure the validity of his interpretation of the conflict in values that characterizes disputes over public lands in the United States.   Whether the issue is building a jetport in the Everglades, damming the Yukon River or excavating an open pit copper mine in the Glacier Peak Wilderness the theme remains the same.   Short-term,

private, financial and speculative values are pitted
against long-term, general public, conservation and
aesthetic values. Ambitious government agencies, arrogant
corporations or wealthy real estate developers propose
environmentally disastrous ventures and conceal inadequate
or nonexistent study and research by mounting public
relations campaigns. The ultimate source of this "unholy
wedlock of the booster and the engineer" is the abuse of
the concept of property. "Land is treated like a commodity when it is in fact a trust." The new land ethic that
Brooks sees emerging finds its "purist expression in the
value attached to our remaining unspoiled wilderness."

29   Broome, Harvey.   "Origins of the Wilderness Society."
       The Living Wilderness 5.5 (July 1940): 13-14+.
A recollection of the events that led to the founding of
the Wilderness Society by one of the founders. Harvey
Broome emphasizes the major role played by Robert Marshall
both in preparing the ground through his article, "The
Problem of Wilderness," and in the actual organization of
the society. Marshall's enthusiasm and his influence on
his many friends and acquaintances assured the success of
the enterprise.

30   Cahn, Robert. *The Fight to Save Wild Alaska.*   New York:
       National Audubon Society, 1982.
An historical overview of the decade long campaign to
assure protection for the Alaskan wilderness. Robert Cahn
opens his account with a portrait of the young, idealistic
lobbyists of the Alaska Coalition on the day in July 1980
when H.R. 39, the Alaska National Interest Lands Conservation Act, would finally be debated on the Senate floor.
He then backtracks to trace early efforts to protect
Alaska's primeval lands and wildlife and to detail
important political events in the 1970s, beginning with
the formation of the Coalition as a cooperating alliance
of conservation organizations and labor unions.   This
dedicated group enlisted the grass-roots support of
citizens from all over the country.   In 1977 school
teachers, students, truck drivers, small businessmen and
homemakers crowded hearings held in places as diverse as
Chicago, Atlanta and Denver to let Congress know that they
valued Alaska as the last American wilderness and wanted
it preserved for their children.   Still, it was another
three years before a bill was signed, and even then the
elation of victory was muted by the knowledge of concessions made to placate Alaska's Senator Ted Stevens.   The
success of the Alaska Coalition signaled a raising of the
nation's environmental consciousness, but Coalition
members viewed the Act not so much as an end as the
beginning of an ongoing effort to ensure its implementation and to correct loopholes through further legislation.

31   Chase, Alston.   *Playing God in Yellowstone: The
       Destruction of America's First National Park.*   San
       Diego, CA: Harcourt, Brace Jovanovich, 1987.
A muckraking analysis of the mismanagement of wildlife and
habitat in Yellowstone National Park based on historical
documents and communications with researchers and rangers.

Alston Chase characterizes the present condition of the park as biologically degraded. Willow and aspen are depleted, beaver and deer are nearly gone, grizzlies are endangered, wolf and other large predators have been eradicated, and the elk herd continues to multiply--all the result of National Park Service officials and environmentalists placing public relations and politics, computer modeling, comfortable myth and abstract metaphysics above field research and reality. Although Chase soundly trounces the earlier park policy of eliminating predators, he reserves his most scathing criticism for the current policy of noninterference and relying on scientifically questionable theories of natural regulation to bring about a balance between elk and forage. It is an illusion that the wilderness ecosystem is intact. In the absence of predators and with the changes brought about by human settlement in the larger ecosystem surrounding the park, it is impossible to maintain Yellowstone as an example of primitive America without active and nurturing restoration by a reformed National Park Service. Chase concludes his book with a twelve point process for reform.

32   Clary, David. *Timber and the Forest Service*. Lawrence, KA: University Press of Kansas, 1986.
A detailed and documented history of the Forest Service from its founding through the 1970s. The author's thorough treatment of the code of the Service explains its otherwise problematical conflicts with environmental groups since World War II. Clary demonstrates with repeated examples that these conflicts have arisen not because the Service has declined from its original mission but because it has carried out that mission as if the country in the last quarter of the 20th century were no different than it had been when Gifford Pinchot preached resource conservation in the Progressive Era. The Forest Service began as an agency dedicated to sustainable yield forestry controlled by silviculture technicians, and it remains so. In spite of the rhetoric of multiple use, foresters find it difficult to conceive of wilderness as a legitimate use of the National Forest System. Concessions to the change in the national culture--hiring wildlife biologists and landscape architects--have served only to refine commodity production. So far, at least, the Service appears unable to seriously dedicate itself to the nonmaterial values of wilderness, scenery and nongame wildlife.

33   Cohen, Michael P. *The History of the Sierra Club*. San Francisco, CA: Sierra Club Books, 1988.
The story of the first eight decades of the Sierra Club told from the points of view of the club's insiders and based on interviews and unpublished documents, as well as on the proceedings of the Biennial Wilderness Conferences and articles in the *Sierra Club Bulletin*. Michael Cohen, the author of an earlier biography of the club's founder, John Muir, traces the evolution of the club from a small group of mountaineers dedicated to preserving and providing access to the Sierras for recreation to a national organization of diverse membership pursuing a variety of

environmental causes. The process of change, as Cohen
makes clear, was a turbulent one involving increasingly
bitter conflicts among club factions and straining the
comfortable relationship the club had once enjoyed with
the Park Service and the Forest Service. Cohen treats his
sources with careful impartiality, especially in interpreting the controversy over David Brower's tenure as
executive director. He is enlightening in showing how the
wilderness conferences both engendered and reflected
changes in the club's philosophy and influence. In the
early 1950s the club disassociated itself from the Park
Service's view of mass recreation, and by the 1969
conference its focus had shifted to the ecological well-
being of the parks and the environmental threats to the
planet.

34   Demars, Stanford E. *The Tourist in Yosemite, 1855-1985:*
          Changing Perceptions of American Wilderness. Salt
          Lake City, UT: University of Utah press, 1991.
A detailed historical interpretation of Yosemite National
Park as a cultural institution that has reflected over
time the perceptions of the American people about the
values to be found in wild nature. Stanford Demars shows
Yosemite as changing its function in response to both
variations in the philosophy of park management and new
developments in recreational taste and technology.
Because of its popularity and tradition of developed
tourist attractions, Yosemite focuses the contrast between
the idea of parks as playgrounds for people and parks as
preserves of pristine America that has dominated thinking
about the national parks since World War II. In the 1960s
and 1970s an articulate wilderness movement succeeded in
bringing about a shift in the balance of park values
towards appreciation and preservation of wilderness.
Backpacking and climbing have increased in popularity,
automobiles are now banned from certain congested areas,
golf and other recreational activities unrelated to a
wilderness experience have been de-emphasized, and nature
interpretation, at least by some park Service employees,
has come to center on the destructive effects of human
activities on nature. In his Epilogue the author suggests
that, although most of these changes are beneficial,
"there is room for concern that the Service might be
neglecting its responsibility to manage the parks for *all*
Americans." Demars book provides an interesting comple-
ment to Runte's recent work on the subject of Yosemite.

35   Disilvestro, Roger L. *The Endangered Kingdom: The*
          Struggle to Save America's Wildlife. New York:
          Wiley, 1989.
A highly readable historical analysis of American wildlife, the development of wildlife law and the changes in
the way Americans have viewed wildlife since colonial
times. The book covers game animals, endangered species
and nongame wildlife in separate sections. Chapters
within each section focus on particular species, the gray
wolf or the pronghorn antelope, for example, or on groups
of related species, such as, waterfowl or bats. Roger
Disilvestro portrays each animal or group as it existed

when the first settlers intruded on its habitat, then proceeds to chronicle its decline under the pressures of settlement and to review the early and ongoing efforts to preserve it. The overall effect of this repeated pattern is cumulative and reinforces Disilvestro's major generalizations about the critical importance of habitat in saving wildlife and about the ambivalent and sometimes pernicious effects of the financial dependence of state wildlife agencies on hunting. Disilvestro is not antihunting; but he makes clear that, for a still dominant segment of wildlife managers and hunters, conservation departments exist to provide large numbers of game animals for hunters to shoot.

36   Ervin, Keith. *Fragile Majesty: The Battle for North America's Last Great Forest.* Seattle, WA: Mountaineers, 1989.
An investigation of the controversy over logging the publicly owned old-growth forests of the Pacific Northwest. Keith Ervin records the opinions and hopes and fears of loggers, environmentalists, industry representatives and Forest Service personnel in order to "delineate the values of old-growth forest, to identify conflicts over their management, and to explore creative solutions." He finds lack of knowledge about the complexity and uniqueness of ancient forests and little constructive dialogue about alternatives to their destruction. Most loggers and Forest Service spokespersons see the big trees as a crop to be harvested. Mature trees are decadent, and leaving them to stand uncut is waste. Supporters of preservation emphasize the role of these forests in stabilizing the climate and protecting biodiversity but also express awe at their majesty and antiquity. This conflict of values is exacerbated by economic pressure from timber companies whose earlier practices of rapid logging on private lands have left them without a steady supply of timber. Ervin concludes that consensus solutions are possible but strong political leadership will be needed to hammer them out.

37   Fox, Stephen. *John Muir and His Legacy: The American Conservation Movement.* Boston, MA: Little, Brown, 1981.
A remarkable history of the conservation movement, emphasizing not so much the overt legal and political developments as the gradual intellectual triumph of Muir's nature preservationism over the resource husbandry espoused by Pinchot. The author introduces the notion of the radical amateur as the driving force leading to the creation of politically powerful environmental organizations. These radical amateurs share certain characteristics. They come overwhelmingly from Anglo-Saxon Protestant backgrounds. Most encountered wilderness or wildlife at an early age. Several knew periods of illness or loneliness as children. A midlife conversion experience is common, in which an earlier acceptance of or even fascination with progress, modernity or technology is replaced by belief in the sanctity of the natural world and the realization that man cannot exist except as a part

of the community of living things.   In a provocative
concluding chapter, Fox argues that conservationism is a
new religion and that it is incompatible with traditional
human-centered monotheisms.

38   Fradkin, Philip L.   *A River No More: The Colorado River
          and The West*.   New York: Knopf, 1981.
The story of the exploitation of the West as it is
reflected in the transformation of the Colorado from a
free-flowing river rushing to the Gulf of California to
a controlled waterway disappearing in a ditch in the
desert.   The author weaves information from his exhaustive
research in public documents and private papers with
quotations from contemporary westerners and evidence from
his own explorations to probe the mythology of reclamation
and the convoluted political alignments that have, until
recently, guaranteed Congressional approval of almost
every proposed water project.   Wherever he visits, as he
traces the course of the river, he uncovers the almost
religious fervor of belief in the efficacy of reclamation.
Reclamation is the proof of man's ability to manipulate
and control nature, as well as to create wealth.
Nevertheless, the diminishing water available for
development and the concern expressed since the 1970s
about the environmental impact of massive water projects
represent the first tentative questionings of the West's
ability to support continued growth.   Fradkin's studies
lead him to conclude that there is a "point, which is
fairly imminent, where the costs greatly exceed the
benefits, and someone has to say, No more."

39   Fradkin, Philip L.   *Sagebrush Country: Land and the
          American West*.   New York: Knopf, 1989.
A companion volume to the author's natural resource
history of the Colorado River and water in the West.   Here
Philip Fradkin intersperses incidents from his own
investigations backpacking and river-running with information from documentary sources to tell the story of the
western grazing lands, especially the Uinta Mountains and
the surrounding sagebrush plains, from geological times
to the present.   The first chapter recounts the foredoomed
efforts of a recent Park Service administrator to change
the name of Dinosaur National Monument to Dinosaur
National Park.   His proposal is effectively and ignomini-
ously defeated by local ranchers and businessmen, who
would welcome the tourist trade park status might bring
but fear further interference with grazing and energy
projects.   This event epitomizes three recurring western
themes: local control of federal regulators, value
conflict between the poles of preservation and private
property rights, and cyclic extraction of resources for
outside markets.   The last chapter traces the events
leading to wilderness status for some of Utah's wild
lands, a victory for preservation in a state influenced
by Mormon notions of dominance and use.   Yet Fradkin
points out an underlying irony.   Whatever the name, this
overgrazed and degraded land tilts toward "impending
disaster." It is a "diseased wilderness."

40   Frome, Michael. *Battle for the Wilderness*. New York:
         Praeger, 1974.
      A broad overview of the history and value of wilderness
      preservation. The author, a frequent contributor to
      environmental journals, recounts the events that led to
      the passage of the Wilderness Act; but he is most infor-
      mative in detailing the long and continuing struggle of
      conservation groups and citizens to force the Forest
      Service to comply with the provisions of the act. Frome's
      emphasis throughout is on the necessity to develop a
      national consensus about the need to limit human consump-
      tion and population growth in order to pass on to our
      descendants a world worth living in. Ultimately, only a
      profound cultural change can permanently protect wilder-
      ness.

41   Gladden, James N. *The Boundary Waters Canoe Area:
         Wilderness Values and Motorized Recreation*. Ames,
         IA: Iowa State University Press, 1990.
      A detailed and scrupulously objective history of the
      controversy over motorized recreation in the BWCA after
      it was designated a wilderness area in the Wilderness Act
      of 1964. A special section of that act, section 4(d)(5),
      permitted continued use of established motorboat routes
      in the canoe country. This anomaly together with improved
      recreation technology and greater use of the area by more
      people made further congressional action inevitable.
      Canoeists, skiers, wilderness advocates and many disinter-
      ested citizens considered mechanized travel incompatible
      with wilderness status and destructive of wilderness
      experience. Many Northeastern Minnesotans, on the other
      hand, viewed any attempt to restrict motor use, including
      the rapidly increasing use of snowmobiles, as an infringe-
      ment on their individual liberties and a threat to the
      local economy. James Gladden presents the acrimonious
      struggle between these factions as a conflict of values
      during a time of "change in social attitudes toward the
      environment." Their differences were not simply a matter
      of personal tastes in recreation but a fundamental clash
      between biocentric and anthropocentric ways of relating
      to the natural world. The legislative outcome of this
      confrontation was a law passed by Congress in 1978, P. L.
      95-495, mandating a significant reduction in snowmobile
      and motorboat use. The appendices contain a helpful
      chronology and a reprint of the law.

42   Graber, Linda H. *Wilderness as Sacred Space*. Washington,
         DC: Association of American Geographers, 1976.
      A review of the literature of wilderness appreciation and
      wilderness recreation for the purpose of determining the
      basic ideas of the wilderness ethic, as it is held by
      "wilderness purists," and of analyzing the role of these
      ideas in "giving form and definition to human encounters
      with nature." Linda Graber argues that the imagery of both
      wilderness writing and wilderness art and photography
      suggests that purists regard wilderness as sacred space.
      They are, in essence, religious devotees who seek to
      transcend the "ordinary world, self, and manner of
      perception" in a numinous wilderness setting. For

purists, wilderness also represents a perfection of ecological order that contrasts sharply with the tamed environment of city and suburb. Graber interprets their almost religious dedication to wilderness preservation as the result of their belief that wilderness is the manifestation of the divine "Wholly Other" or Absolute. Their commitment, in turn, enables them to set the terms of the environmental debate and to influence legislation and political action beyond their numerical strength. According to the author, this book is a tentative study. Certainly, some of its conclusions would be disputed by other writers on wilderness values.

43   Graf, William L. *Wilderness Preservation and the Sagebrush Rebellions.* Savage, MD: Rowman and Littlefield, 1990.
An examination of how the evolution of federal land policy over the past century has influenced the development of a national wilderness system. Geographer William Graf's knowledge of the arid West provides a scientific context for his detailed treatment of the recurring cycles of explosive development, government regulation and western reaction that have resulted over time in transforming a land policy of disposal to one of preservation. The first sagebrush rebellion, formed to defeat John Wesley Powell's plan for limited and orderly development of the West based on scientific research, was successful. The failure of later rebellions--against President Theodore Roosevelt's forest reserve policies, against the restrictions imposed on graziers by the Taylor Grazing Act during the 1930s and 1940s, and in opposition to wilderness classification of BLM lands in the late 1970s and under President Reagan--has firmly established the National Wilderness Preservation System as a unique contribution of American culture. Though the perennial conflict between commodity users and preservationists will, in one way or another, continue, Americans today have the "opportunity to recreate one of the country's most basic historic and geographic experiences: to be alone in the wilderness." In addition to its treatment of overall historical themes, Graf's book presents a collection of succinct portraits of the many individuals who have been active and influential in political conflicts over the use of the public lands.

44   Graham, Frank, Jr. *Man's Dominion: The Story of Conservation in America.* New York: Evans, 1971.
A history of the conservation movement from the 1880s to the passage of the Wilderness Act in 1964. Frank Graham, Jr. writes in lively, forthright prose to give non-specialist readers the historical background necessary for understanding the environmental controversies that were taking shape in the 1970s. Because he relies heavily on the actual words of famous conservationists and their opponents to tell his story, his narrative has an immediacy that brings to life past controversies and characters. There is no question as to where Graham's loyalties lie. He writes as a dedicated conservationist whose voluminous reading has only confirmed the venality of the those who fought against wilderness preservation.

45   Haag, John. "A Yearning for Synthesis: Organic Thought
            Since 1945."   *International Journal of Social
            Economics* 8.7 (1981): 87-111.
      A survey of intellectual and social history since 1945
      that places changes in the concept of the relationship
      between man and nature within the context of an emerging
      new cultural norm that will also involve changes in man's
      relationships with God and with social institutions. John
      Haag argues that the events of World War II brought an end
      to the credibility of the idea, dominant for several
      centuries,   that   rationality   and   mechanistic   science
      assured "human perfectibility and unlimited economic
      growth." This has led to "vigorous rethinking of a number
      of fundamental Western beliefs."   The organic view of
      nature that characterizes ecological thinking has begun
      to replace the atomistic view of mechanistic science.   In
      the organic view humans are part of an interconnected
      whole, in which events are not isolated and technology
      used to solve a specific problem can have multiple and
      unforseen effects. Other creatures within this whole are
      entitled to their existence, and we have no mandate to
      exterminate them. The emerging norm recognizes religious
      attitudes as being as necessary as reason in answering
      ultimate questions. It seeks to replace satisfaction of
      individual material needs with building livable communi-
      ties as the goal of society. Haag does not treat wilder-
      ness values directly; nevertheless, his article should be
      read for an understanding of how these fit into what he
      sees as the task of the late twentieth century: "the
      creation of a new culture complete with radically new
      values and attitudes."

46   Hays, Samuel P., and Barbara D. Hays. *Beauty, Health,
            and Permanence: Environmental Politics in the United
            States, 1955-1985.* New York: Cambridge University
            Press, 1987.
      A political history of the environmental movement in the
      United States, covering wilderness issues as well as anti-
      pollution campaigns and concern over air and water
      quality.   The authors contend that all facets of the
      increasing public concern over the environment are the
      result of a deep-seated change in values during the period
      of rising affluence, urbanization and educational attain-
      ment after World War II. Americans have come to view the
      good life as including opportunities for wilderness
      experiences and expect clean water and air. The depth of
      this value transformation is shown by the failure of the
      Reagan administration's anti-environmental efforts to
      reverse the trend of the previous twenty years.

47   Huth, Hans. *Nature and the American: Three Centuries
            of Changing Attitudes.* Lincoln, NB: University of
            Nebraska Press, 1957, 1972.
      A survey of the varied attitudes Americans have held
      toward wild nature since the seventeenth century designed
      to show that conservation is rooted in the American
      tradition.   The author reviews the work of writers,
      painters and explorers to illustrate that as early as the
      eighteenth century a significant minority of Americans

reacted positively to the aesthetic qualities of natural
scenery, sometimes contrasting the beauties of wilderness
with the crowding and odors of the rapidly growing cities.
Until after the Civil War conflicting and ambivalent
attitudes toward wild nature could exist amicably side by
side among different groups or within the same person at
different times. The clash between conservationists and
advocates of unrestrained growth came to dominate our
politics and our perceptions only after the excesses of
the late nineteenth century had clearly demonstrated the
power of human activities to exhaust and degrade the land.
Huth makes no attempt to cover individuals or political
events in detail; the contribution of this book lies in
compiling literary and artistic evidence for a continuing
tradition of aesthetic appreciation of nature in American
culture.

48   Limerick, Patricia Nelson. *Desert Passages: Encounters
         with the American Deserts.* Albuquerque, NM:
         University of New Mexico Press, 1985.
A study of human attitudes toward wild nature in the form
of concrete, actual places in the American deserts.
Patricia Limerick analyzes the works of eight writers
about the desert, from John C. Fremont's efficient travel
records in the 1840s to Edward Abbey's contemporary
celebration of desert sparseness and simplicity, in order
to identify general attitude changes within the context
of complex individual encounters with the reality of the
arid lands of the West. Responses to the desert diverge
from the recognized developmental sequence of wilderness
attitudes, from fear and hatred to mastery of resources
to aesthetic appreciation, in that mastery over the desert
is never securely achieved. The desert, even dammed and
irrigated, remains a place inimical to the biological
processes of human life. Indeed, modern appreciation of
the desert, as it is expressed by Abbey and, to some
extent, by Joseph Wood Krutch, rejoices in the desert's
contrast to a crowded, increasingly urban and manipulated
world. The harsh, dry climate sets limits that humans
only postpone by mining groundwater. Today, in spite of
technological triumphs, the desert continues to threaten
the future of resource gambling and treasure hunting with
its uncertain ability to sustain heavy settlement and use.
Limerick's book enriches more general studies of attitude
change.

49   Manes, Christopher. *Green Rage: Radical Environmentalism
         and the Unmaking of Civilization.* Boston, MA:
         Little, Brown, 1990.
A historical explanation of the emergence and significance
of radical environmentalism, particularly, ecotage,
monkeywrenching and civil disobedience, in defense of
wilderness. Author Christopher Manes, a former associate
editor of the journal *Earth First!*, is not a neutral
reporter; he fully supports efforts to "stop the culture
of technology from unraveling the fragile, resplendent web
of life on this planet." Nevertheless, his is a thorough-
ly documented, widely researched and clearly organized
book. In addition to summarizing major confrontations

between members of Earth First! and other radical groups
and the employees and officials of timber, mining and
whaling corporations, Manes interprets the emergence of
these groups in the 1970s as the response of people with
a deep emotional sensitivity to ecological destruction to
the increasing dominance of mainstream conservation
organizations by considerations of career and compromise.
Radical environmentalists share the world picture of deep
ecology--the conviction that humankind is not the center
of value, that other species have as much right to exist
as humans do and that "wilderness not civilization is the
real world." They are engaged in a last ditch attempt to
deflect "progressive conversion of the natural world into
cultural artifacts measured by industrial output." The
preservation, even restoration, of wilderness lies at the
heart of their program because wilderness is the true home
of all life. Once wilderness is gone there is no reference point outside the manipulations of culture, state and
other powerful interests. These are then free to shape
citizens as they will. Manes book is an essential source
for understanding contemporary wilderness values.

50   Martin, Russell. *A Story That Stands Like a Dam: Glen
        Canyon and the Struggle for the Soul of the West.*
        New York: Holt, 1989.
The history of Glen Canyon Dam told with colorful detail
and attention to critical incidents and personalities.
Separate chapters concentrate on different aspects of the
Glen Canyon project. This technique enables the author
to convey the excitement and enthusiasm of the engineers
and contractors who built this monumental structure and
the dedicated emergency salvage operations of the archae-
ologists who worked at the Anasazi sites in the canyon,
as well as the thoughts and opinions of politicians,
conservationists and river runners. Overall, Russell
Martin sees the Glen Canyon story not only as a conflict
at a particular time and place between those who expected
the dam to bring wealth and comfort to the arid West and
those who deplored the unnecessary and heedless destruction of incomparable natural beauty but as a turning point
in the history of American environmental values. In the
depression years of the 1930s the Hoover Dam was almost
universally acclaimed as a triumph of technology and the
human spirit. In the 1960s, as Glen Canyon filled to
become Lake Powell, the mood was far more uncertain.
After Glen Canyon public opinion could no longer be
counted on to support further damming of the Colorado; and
the Bureau of Reclamation, for thirty years a symbol of
engineering sophistication and technical expertise, began
a period of slow decline in power and prestige.

51   Matthiessen, Peter. *Wildlife in America.* Revised
        edition. New York: Viking, 1987.
An overview of the history and contemporary status of
American wildlife organized in chapters devoted to
particular habitats or ecosystems. Each chapter gives an
account of wildlife species and populations as they
appeared to explorers and settlers, traces their decline
or, in some cases extinction, under the pressures of civi-

lization, and assesses their current condition and the
effectiveness of the laws and conservation measures meant
to protect them.   Peter Matthiessen writes in a clear,
mildly satiric style: yet this is a gut-wrenching tale of
wanton destruction, greed, waste and ignorance continuing
even in the 1980s because political pressures and expedi-
encies   control   government   conservation   agencies   and
interfere with law enforcement. Meanwhile the "accelerating downward spiral of life's diversity throughout North
America" goes on. Matthiessen finds hope, however, in the
increasing interest of ordinary Americans in preserving
wild habitats--the "reaction of a people entrapped by the
apparatus of its own progress, and seeking a passage back
to more permanent values, to the clean light of open air."

52   McCloskey, Michael.   "The Wilderness Act of 1964: Its
            Background and Its Meaning." *Oregon Law Review* 45
            (1966): 288-321.
A legal history and analysis of the Wilderness Act by a
practicing   lawyer   and   Sierra   Club   activist.   Michael
McCloskey attributes the emergence of this law in the
1960s to a combination of the increasing scarcity of
wilderness areas with the existence of economic surpluses
and highly educated leaders cognizant of the role of
wilderness in American cultural history. His background
discussion includes a summary of typical wilderness values
as well as a legislative history, which briefly records
Congressional actions from 1956 through the passage of the
final bill.   The bulk of the article probes the language
of the Wilderness Act and points out ambiguities that
would require resolution through judicial interpretations.

53   McCloskey,   Michael.   "Wilderness   Movement   at   the
            Crossroads, 1945-1970." *Pacific Historical Review*
            41 (1972): 346-361.
A paper by the then Executive Director of the Sierra Club
analyzing the stages of the wilderness movement after
World War II.   The author sees an early, primarily
defensive   stage   followed   by   the   offensive   campaign
culminating in the Wilderness Act of 1964.   The third
stage, which has left the movement uncertain and dissipat-
ed, results from the constant and continuing effort needed
to satisfy the provision of the act that requires an
affirmative vote of Congress on each addition to the
wilderness system and from loss of fervor as the wilderness movement is subsumed in the larger environmental
movement.

54   McConnell, Grant. "The Conservation Movement: Past and
            Present." *Western Political Quarterly* 7 (1954):
            463-478.
A convincing interpretation of the state of the conservation movement at mid-century based on an analysis of the
tension   between   the   Pinchot   and   Muir   strains   in   its
historical development.   Grant McConnell contrasts the
"small, divided and frequently uncertain" movement as it
existed in the early 1950s with the major political
strength of conservation as the natural resources compo-
nent of Progressivism.   In the first decades of the

century conservation meant the use of our natural resources for the benefit of the greatest number over longest
time. It appealed to the doctrine of equality and was
linked with popular social and economic reforms. To be
against conservation was to be selfish in the way of large
corporations and monopolies. Reality, however, proved too
complex for easy decisions based on a utilitarian formula
that fails to distinguish among uses except as these
appeal to more or fewer people. "Given the character of
American society this amounts to an emphasis on material
values." The groups that makes up the new conservation
movement share the conviction that noncommercial uses of
natural resources are important and that some uses,
wilderness, for example, are superior to economic considerations. At the root of the new conservation is the
belief that man has an obligation to the natural environment.

55   McDonald, Corry. *The Dilemma of Wilderness*. Santa Fe,
         NM: Sunstone Press, 1987.
Reminiscences of the role of wilderness and wildlife
organizations in defining and setting aside wilderness
areas in New Mexico since the 1960s. Corry McDonald was
an active member of the New Mexico Wilderness Study
Committee, an overview organization made up of representatives of a wide spectrum of conservation and outdoor
organizations. He recounts from personal experience the
ongoing involvement and hard work of dedicated laypersons
in gathering factual data and pressuring the Forest
Service, the National Park Service and the Bureau of Land
Management to carry out their obligations under the
Wilderness Act. It proved to be nearly a full-time
activity, at least, for the leadership. The work of his
organization did not end with the enactment of legal
wilderness areas. The dilemma of wilderness reappears in
the debate over wilderness management. Too many visitors
can destroy wilderness values, and extensions of the
economy of civilization encroach into the wilderness in
the form of incompatible uses--ORVs, water diversion,
grazing and mining. The appendix includes documents
presenting Atlantic Richfield's concerns about locking up
minerals and oil and gas in wilderness areas as well as
McDonald's own rebuttal of the company's economic estimates.

56   Merchant, Carolyn. *Ecological Revolutions: Nature,
         Gender, and Science in New England*. Chapel Hill,
         NC: University of North Carolina Press, 1989.
An analysis of New England history in terms of ecological
revolutions, the "processes through which different
societies change their relationship to nature." Carolyn
Merchant examines three historical social constructions
of nature in New England together with their concomitant
modes of production, reproduction and consciousness:
Native American reciprocity between an active nature and
humans; colonial preoccupation with maintaining a firm
separation between humans and wild nature; and capitalist
or industrial denigration of nature to passive object.
Revolutions in these ways of organizing and experiencing

nature occurred when modes of production and reproduction conflicted with the requirements of ecology in a given habitat. Although the bulk of Merchant's text deals with events prior to the 20th century, she includes an "Epilogue" that speculates on the possibility that we are now undergoing a further ecological revolution, in which humans will once again come to see themselves as strands in a web of active partnerships with the natural world. Such a change might be characterized by a land ethic that bases values on ecosystem viability and mutual obligations rather than on human self-interest and individual economic rights. The main text is accompanied by extensive charts, notes and bibliography.

57   Merchant, Carolyn. *The Death of Nature: Women, Ecology, and the Scientific Revolution.* New York: Harper & Row, 1980.
A feminist history of the change in the Western world picture from organism to machine between 1500 and 1700. Merchant argues convincingly that this change in the way of looking at the world degraded both nature and women and created the political and ideological context for the exploitation and destruction of natural ecosystems that characterize the modern world. Descriptions of nature have a normative impact on how nature is treated. Within that 200 year time span the tension between the earlier metaphor of nature as a living organism and nurturing mother and new activities and circumstances--increasingly commercial agriculture, expanded metallurgy and mining, and population pressure--became too great to sustain. The metaphor of the machine is better suited to the unrestrained exploitation of the earth required in the service of power, technology and wealth. Sustained yield conservation is compatible with this mechanical model. Both feminism and the holism of ecological environmentalists, however, represent a break with the dominant world view and signal the re-emergence of an organic model of reality.

58   Nash, Roderick. "Path to Preservation: An Overview." *Wilderness* 48.165 (Summer 1984): 5-11.
A succinct history of the events leading to the passage of the Wilderness Act remarkable for the clarity with which it conveys the significance of the Act both as legislation and as the institutionalization of a philosophical concept. Prior, to 1964 defined primitive or wilderness areas existed at the behest of federal agency regulations; their status could be converted to resource use at any time by order of the appropriate appointed official. The Wilderness Act created legally sanctioned wilderness areas subject to change only through an act of Congress and formally recognized the enduring value of wilderness to the American people. A special feature of Nash's article is his detailed discussion of Howard Zahniser's pivotal role in devising the language of the legislation and in gathering support for it both in Congress and among the conservation community.

59   Nash, Roderick. *Wilderness and the American Mind.* 3rd
           edition. New Haven, CT: Yale University Press, 1982.
     An indispensable book in the study of wilderness values.
     Nash uses the central thesis that wilderness is a state
     of mind, an idea, rather than a place, to explore how the
     meaning of wilderness has changed throughout American
     history. The breadth of his scholarship and his deep
     familiarity with the literature enable him not only to
     trace the broad outlines of the major trends in attitudes
     toward wilderness from fear and revulsion to appreciation
     and preservation, but also to discriminate the more subtle
     differences in philosophy among wilderness proponents.
     The preface to the third edition examines the idea that
     the concept of wilderness is the product of civilization,
     the change in human society from hunting and gathering to
     herding and farming. Other additions include chapters on
     wilderness philosophy, Alaska, and international develop-
     ments. There is an excellent updated bibliography.

60   Nash, Roderick Frazier, ed. *American Environmentalism:
           Readings in Conservation History.* 3rd edition. New
           York: McGraw-Hill, 1990.
     An anthology of brief readings arranged so as to facili-
     tate an understanding of the changes in the purposes and
     values that have inspired efforts to conserve and protect
     nature since the late nineteenth century. Editor Roderick
     Nash introduces each group of readings with an analysis
     of the significant conservation problems and contributions
     of the historical period it represents. The essays
     themselves illustrate the convictions and concerns of the
     time and cover pollution and overpopulation as well as
     wilderness and species extinction. A section titled "The
     New Environmentalism, 1973-1990" has been added since the
     1976 edition. It provides a spirited introduction to deep
     ecology, monkeywrenching and the tactical split between
     old-line conservation organizations and newer, more
     radical groups, such as, Earth First!. Nash teaches
     environmental studies at the University of California; his
     book could well be used as a text for similar courses
     elsewhere.

61   Nash, Roderick Frazier. *The Rights of Nature: A
           History of Environmental Ethics.* Madison, WI:
           University of Wisconsin Press, 1989.
     A wide-ranging and scholarly history of the shift in the
     moral grounds for environmental protection from resource
     conservation to ecological integrity in the United States
     in the 20th century. Although the author surveys the
     roots of the idea that morality ought to concern the
     relationships of humans to nature from the "jus animalium"
     of the Romans through the long centuries of neglect to the
     works of John Ray in the late 17th century and Henry Salt
     and the humane movement in the late 19th and early 20th
     century, his principal contribution lies in intellectually
     ordering the many-faceted religious, philosophical, and
     nature liberation movements which have flourished since
     the 1960s. Nash sees the profound change in ethics that
     seems to be occurring as being, at the same time, both a
     radical departure from conventional American exploitative

attitudes and a logical and emotional extension of the
American commitment to natural rights. Granting rights
to nonhuman beings and their communities is in the liberal
tradition of the abolition of slavery and, more recently,
the civil rights and women's liberation movements. The
text is well documented, and the appended bibliographic
essay should prove valuable to scholars.

62   Odell, Rice. *Environmental Awakening: The New Revolution
         to Protect the Earth.* Cambridge, MA: Ballinger,
         1980.
An overview and analysis of the significance of the
environmental movement of the 1960s and 1970s based on
material drawn from the *Conservation Foundation Letter.*
The author sees the coming together during that period of
various groups and interests, from wilderness advocates
to public health experts, as the beginning of a true
revolution in cultural values. Underlying the views
expressed by environmentalists of whatever background or
affiliation is a fundamental recognition of the importance
of nature in supporting a healthy, safe, and perceptually
rich human life and a refusal to accept political or
social solutions that degrade the ecosystem for short-term
or private gain. Odell treats different aspects of the
environmental crisis--population, chemicals, energy,
etc.--in separate chapters. Wilderness and wildlife
preservation are covered in the chapter on values, where
he outlines utilitarian and ethical arguments for protecting endangered species and habitats. This broad overview
serves as a useful introduction to environmental history
prior to the Reagan years.

63   Oelschlaeger, Max. *The Idea of Wilderness from Prehistory
         to the Age of Ecology.* New Haven, CT: Yale
         University Press, 1991.
An exciting intellectual history of the concept of
wilderness as it has evolved over time in conjunction with
changes in human culture from hunter-gatherer to urban-
industrial. Max Oelschlaeger approaches his study of the
ideas of wilderness expressed in Paleolithic, ancient,
modern and postmodern world views from the perspective,
heretical to modern thought, that the differences in the
basic cultural forms of these societies do not represent
a simple linear progression from primitive to advanced.
Once we recognize that modern Euroculture is not the norm
for human existence, we can begin to appreciate the
validity of the prehistoric sense of the organic unity and
kinship of humankind with wild nature and to question the
total humanizing of the earth's landscape that is rapidly
taking place as the result of the modern dismissal of
nature as mere environment to be exploited as "resource
to fuel the human project." Oelschlaeger analyzes the
writings of the important postmodern critics of the idea
of wilderness as unnecessary and opposed to civilization.
Separate chapters are devoted to Thoreau, Muir, Leopold,
Robinson Jeffers and Gary Snyder, while a later chapter
surveys contemporary wilderness philosophy. In his final
chapter, he speculates that "humankind may be on the brink
of a postmodern age," in which we will again experience

ourselves as finite beings embedded in the flux of a
sacred cosmos. This difficult but rewarding book is a
major synthesis of insights from fields as diverse as
physics, sociology, religion and philosophy.

64   Palmer, Tim. *Endangered Rivers and the Conservation
         Movement*. Berkeley, CA: University of California
         Press, 1986.
A history of people destroying and protecting rivers. The
author details the outstanding political efforts to
prevent destruction of splendid scenic areas by damming,
including the successful prevention of the Echo Park Dam
on the Green River, which would have flooded Dinosaur
National Monument, and the disheartening failure to stop
the Glen Canyon Dam, which resulted in permanently
drowning the beautiful Glen Canyon of the Colorado.
Equally important, he makes the reader aware of the sheer
volume of dam-building projects already proposed at less
well-known sites. The appended list by state of endangered rivers gives the name of the river, the proposed
project, its sponsor, purpose, and status, with an
additional column for explanatory notes. Palmer sees
increasing recognition by the public of the wildlife
habitat and recreational values of wild rivers as a
significant factor in countering the mounting demand for
dams that will result from future perceptions of the
scarcity of water resources.

65   Palmer, Tim. *The Sierra Nevada: A Mountain Journey*.
         Washington, DC: Island Press, 1988.
A journey through the Sierra Nevada from Quincy in the
north to Tehachapi in the south on the eve of the passage
of the California Wilderness Act in 1984. Tim Palmer
meets and gets to know the diverse people in small towns
and along the trails who devote a portion of their lives
to defending the Sierra wilderness. Through letters,
testimony, protests, conservation activities and teaching,
these people attempt to stop the road and dam building,
cattle grazing and clearcutting that threaten the wild
lands. As he wends his way south, Palmer summarizes in
capsule form the history of each region, outlines its
current conservation controversies and speculates about
its environmental future. Though much of his trip is by
van, he backpacks for 30 odd days in the mountain country
between Ebbetts Pass and Lone Pine, rediscovering his own
wilderness values along the way. The main body of the
book concludes with a report of local environmentalists
celebrating on the day the California bill passes Congress. The author calls this a "tribal gathering" of a
"mass mutiny from civilization" and maybe the "beginning
of a real civilization." His epilogue, however, shows
exploitation still proceeding apace. Several appendices
provide facts about the Sierra Nevada.

66   Palmer, Tim. *Stanislaus: The Struggle for a River*.
         Berkeley, CA: University of California, 1982.
The story of the long and losing fight to save the white
water canyons of the Stanislaus River from burial beneath
the New Melones Reservoir. Tim Palmer, who speaks both

as reporter and participant, opens his book with a light-
hearted account of a rafting trip down the Stanislaus in
July of 1979 before the dam was filled. The following
chapters detail major activities in the Friends of the
River campaigns over a ten year period beginning in 1971
with the Proposition 17 referendum and concluding with the
effort to obtain wild and scenic river designation from
Congress in 1980.   Palmer's deft characterization of
individuals enables him to convey the intensity of the
radical protests against the dam in the late 1970s and the
integrity of Mark Dubois, the hero who chained himself to
the rocks to stop the Army Core of Engineers from filling
the dam.   Although he portrays a classic conservation
conflict between New Melones supporters with a "traditional outlook on control of nature for the benefit of man
through   the   economy,"   and   idealistic   opponents   who
volunteer   their   time   and   their   careers   for   love   of   a
river, he also reveals changes in values that take place
over time.   For protesters, recreation gives way to
protecting remnant wild places; and even some proponents
of the dam come to recognize the need for water conservation as an alternate to increasingly expensive engineering
projects.

67   Runte, Alfred.   *National Parks: The American Experience.*
           2nd   ed.,   revised.   Lincoln,   NB: University   of
           Nebraska Press, 1987.
       A history of the changing rationale for national parks and
       the   resulting   controversies   and   political   maneuvering.
       Originally, the parks were set aside to preserve spectacu-
       lar scenery on what was considered to be commercially
       worthless land.   By the mid-twentieth century, however,
       the   role   of   the   parks   in   protecting   and   preserving
       ecosystems   and   wild   species   was   becoming   increasingly
       important.   Ironically, the breeding grounds and grazing
       lands necessary to fulfill these broader functions have
       historically   been   excluded   from   park   boundaries.   The
       contemporary question, yet to be resolved, is whether we
       as a nation will be willing to enlarge the parks so that
       they become true ecological preserves. The second edition
       adds a chapter on Alaska as the last chance for preserving
       whole ecosystems and an epilogue on the legacy of the
       Reagan administration.

68   Runte,   Alfred.   *Yosemite: Embattled   Wilderness.*
           Lincoln, NB.   University of Nebraska Press, 1990.
       An account of the classic confrontation between preserva-
       tion in the history of Yosemite National Park.
       Alfred Runte shows how the ideal of the park as sanctuary,
       a "vignette of primitive America," has been repeatedly
       compromised by the National Park Service's concern with
       accommodating an increasing number of visitors and by the
       concessionaires' success in convincing park officials and
       politicians   that   money   making   schemes   for   attracting
       tourists represent the fulfillment of legitimate needs.
       In spite of the efforts of Joseph Grinnell and other
       conservationists to introduce the idea of the park as an
       ecological   island   for   wildlife   research   and   the   park
       ranger   as   an   interpreter   of   the   natural   world,   park

management has continued to emphasize entertainment over understanding. The vicious cycle of encouraging bears to feed at garbage platforms for the amusement of visitors followed by shooting them on the least complaint about their aggressiveness finally ended in the 1970s when evidence of mass killings outraged public opinion. Yet the park continues to be managed in a way that degrades biological resources rather than educating visitors to accommodate themselves to the wild and respect the chain of biological relationships that is the ultimate wealth of the wilderness.

69   Schrepfer, Susan R. *The Fight to Save the Redwoods: A History of Environmental Reform.* Madison, WI: University of Wisconsin Press, 1983.
A detailed history of the decades long effort by the Save-the-Redwoods League and the Sierra Club to preserve the California redwoods. The author focuses on the mid-century environmental, social and scientific changes that transformed the environmental preservation movement from polite behind-the-scenes persuasion to militant public advocacy in the fifties and sixties. Three major developments formed the context for this change. Post-war population growth and technological advance increased both the power of the timber industry to rapidly clearcut irreplaceable forests and the demand for lumber. The membership of environmental groups grew in size and changed in composition. Most fundamental, however, were the scientific developments that culminated in the 1950s. Directional theories of evolution, with their accompanying optimism about progress, were discredited. Man was not the end of the process and held no privileged position in nature. The future of wildness was not assured. Conservationists needed to act promptly and radically to save species and ecosystems rather than negotiate to spare a few monumental specimens of the tall trees. Schrepfer's book is an insightful effort to explain a complex social change.

70   Searle, R. Newell. *Saving Quetico-Superior: A Land Set Apart.* St. Paul, MN: Minnesota Historical Society, 1977.
A history to 1977 of the long and herculean effort of dedicated conservationists to obtain wilderness protection for the Quetico-Superior canoe country. It is especially the story of the work of the Superior-Quetico Council and its chief spokesman Ernest Oberholtzer, who formulated the dream of a canoe wilderness in the early 1920's and did not withdraw from the decades long participation until 1970 at the age of 86. In between, the council and other conservation groups faced the powerful dam-building timber baron Edward Backus, as well as a seemingly endless array of plans and actions whose effects would compromise the pristine character of the region. As the author points out, misunderstandings and confusions over the role of the national forests and the definition of a wilderness area exacerbated continuing conflicts even among supporters. Ironically, though the Quetico-Superior was one of the

earliest areas to become the focus of wilderness preservation, it has yet to eliminate all nonconforming uses.

71   Tober, James A.   *Wildlife and the Public Interest:
        Nonprofit Organizations and Federal Wildlife Policy.*
        New York: Praeger, 1989.
An informed analysis of the role of nonprofit organizations in bringing about changes in the institutionalized
model of human-wildlife interaction and in federal
wildlife policy since midcentury. Tober concentrates on
detailing the methods--lobbying, testifying, litigation,
education, and marketing--that these organizations employ
to influence laws, administrative procedures, and the
attitudes of elected officials and government administra-
tors. His wide reading and frequent interviews with
individuals from various associations and agencies enable
him to detail specific cases of government and nonprofit
organization interaction, especially, the attempt to save
the California condor and the challenge to exporting
bobcat pelts. Each of these is covered in a separate
chapter. Environmental activists will find Tober's
assessment of what works and what does not work fascinating reading; however, the author's somewhat discursive
style hinders a full articulation of the degree of overall
effectiveness of nonprofit organizations in steering the
nation through a major change in the values that govern
wildlife decision-making.

72   Trevelyan, G. M.   "The Call and Claims of Natural
        Beauty."   *Voices for the Wilderness.*   Ed. William
        Schwarz.   New York: Ballantine, 1969: 143-160.
A percipient analysis using literary references of the
change in the appreciation of wilderness scenery at the
time of the Industrial Revolution. The author argues that
this was the result of the degradation of the rural
countryside. Human beings have always responded to the
combination of aesthetic value, the promise of life and
renewal, and the sense of belonging on the earth that
makes up the experience of beauty in nature; but, for most
people today that beauty can be found nowhere but in the
remote wildernesses. This essay was originally published
in the Proceedings of the 6th Wilderness Conference
entitled *The Meaning of Wilderness to Science.*

73   Turner, Frederick.   *Beyond Geography: The Western
        Spirit against the Wilderness.*   New Brunswick, NJ:
        Rutgers University Press, 1983.
An interpretation of Western history written to answer the
author's nagging question of why he and other Americans
remain, even today, ignorant of and estranged from the
American land. The explanation Frederick Turner arrives
at through analysis of works on the beginnings of civilization in the Near East and of the codification of
Christianity during the late Roman Empire is a spiritual
one. The Israelites substituted sacred history for sacred
myth, and Christianity went further to seal off further
revelation after the unique historical event of the coming
of Christ. The "older, organically derived feelings of
gratitude towards nature and of the vital interdependence

of all things" are supplanted by an "enduring opposition
of man and nature." Cut off from contentment in the
natural world, Christians sought in repeated outbursts of
aggression against heathen and heretic to maintain the
vain hope of winning a lost paradise. The frenzied
destruction and denigration of the American wilderness and
the despicable treatment of Native Americans have their
roots in our view of nature as without spiritual significance. Turner's rendering of encounters between whites
and Indians from Columbus to the hysterical reaction to
the Sioux ghost dancers in the 1890s goes far to dispel
the comfortable illusion that Western man triumphed
because of his superior culture.

74   Udall, Stewart. *The Quiet Crisis*. New York: Holt,
        Rinehart and Winston, 1963.
A short but politically significant history of the
relationship of the American people to the American land
published while the author was Secretary of the Interior
under President Kennedy. The introduction, written by
President Kennedy, echoes Udall's historical interpretation and calls for a commitment by all Americans to
preserving the land and conserving resources. The
author's familiarity with politics allows him to make a
telling comparison between the resource raiders of the
19th century and their more sophisticated 20th century
brethren. The raiders of the late 19th century used the
myth of superabundance to make their fortunes at the
expense of destruction of resources for the future; the
new raiders exploit the rationale of the technological fix
to accomplish the same selfish purpose. This is an
important book for understanding the climate of opinion
that made passage of the Wilderness Act possible.

75   Udall, Stewart. *The Quiet Crisis and the Next
        Generation*. Salt Lake City, UT: Peregrine Smith,
        1988.
An update to the author's 1963 book, *The Quiet Crisis*.
This new volume reprints the original and adds chapters
covering the period since 1962. Udall treats Rachel
Carson, David Brower and Howard Zahniser separately as
three seminal figures in the environmental movement.
Through her books, Carson introduced the ordinary concerned citizen to the concepts of ecology; Brower's
fearless zest for combat and skill in mass marketing built
widespread support for wilderness preservation; and Howard
Zahniser, through his patience, political savvy and good
humor, brought about a wilderness constituency in Congress. Udall's tribute to Zahnie, as he was called, is
especially moving. In the final chapter the author
summarizes the achievements of the "sophisticated,
assertive environmental movement" that has emerged since
Silent Spring.

76   Watkins, T. H, and Charles Watson, Jr. *The Land No One
        Knows*. San Francisco, CA: Sierra Club Books, 1975.
An impassioned, sometimes satiric, popular history of the
profligate way in which the public domain lands have been
subject to speculation, give-away, raiding and over-

grazing since the beginning of the Republic. The first
section, called "The Inheritance," concludes with the
effort of stockmen in the 1940's to regain their control
over public grazing lands that had been modified by the
provisions of the Taylor Grazing Act of 1934. The second
part documents the failure of the doctrine of multiple
use, as interpreted by the Forest Service and the Bureau
of Land Management, to protect the wilderness values of
natural vegetation and wildlife against the political
pressure of ranchers, oil-men, and motorized recreation-
ists. The appendices include a chronology of major public
land laws through 1964, as well as an index of wildernesses within the BLM lands.

77   White, Lynn, Jr.   "The Historical Roots of Our Ecologic
        Crisis." *Science* 155 (10 March 1967): 1203-1207.
      A short article that is often referred to in later and
      more extensive explanations of modern environmental
      degradation as the result of traditional Western ideas
      about   the   relationship   between   humans   and   nature.
      According to historian Lynn White, Jr., our present
      situation stems from the immense power generated by the
      fusion of science and technology that took place around
      1850; but these tools of Western success in controlling
      nature have their roots in Christianity, "the most
      anthropocentric religion the world has seen."   To keep
      ourselves from destroying the planet we need to abandon
      the Christian notion that nature exists to serve man and
      can be exploited with indifference to the biosphere.   For
      practicing Christians, among whom White counts himself,
      he suggests reviving the alternate view of nature taught
      by St. Francis.

78   Worster,   Donald.   *Nature's Economy: A History of
        Ecological   Ideas*.   Cambridge,   MA: Cambridge
        University Press, 1985.
      An absorbing account of how the science of ecology has
      influenced and been influenced by cultural values and
      ruling metaphors. Donald Worster portrays Western thought
      since the 18th century as vacillating between two contra-
      dictory ways of reasoning about nature: a search for value
      and ultimate order and an opposite drive to dominate the
      natural world.   These points of view are encapsulated in
      metaphors of organism or machine or factory.   Worster
      terms those who find nature an organic community "arcadi-
      ans" and identifies Gilbert White as their precursor and
      Thoreau, Darwin, Frederic Clements and, more recently,
      William Wheeler and Charles Elton as spokesmen for their
      conviction that mankind must learn to accommodate itself
      to the natural order.   The opposite ecological strain,
      which urges imperial control and management of the nature
      factory for human purposes, can be traced from Linnaeus
      through T. H. Huxley and other Victorians to Arthur
      Tansley and the New Ecologists.   Given this historical
      ambivalence of ecology towards man's domination of nature,
      Worster argues that, however enlightening the science may
      be in increasing our knowledge of the world, it cannot
      serve as the unequivocal basis for attributing value to

wild nature. This compelling and provocative book should
be included in any study of contemporary ecophilosophy.

79   Zaslowsky, Dyan, and The Wilderness Society. *These*
        American Lands: Parks, Wilderness, and the Public
        Lands. New York: Holt, 1986.
      A clear, coherent survey of all the American public lands,
      including the National Park System, the National Forest
      System, the National Wilderness Preservation System, The
      National Resource Lands of the BLM, the National Wildlife
      Refuge System, the National Interest Lands of Alaska, and
      the National Wild and Scenic Rivers and the National
      Trails Systems. Separate chapters trace the historical
      development and outline agendas for the future of each
      designated land group. One advantage of this treatment
      is the opportunity it provides for the incremental
      repetition of certain themes and values. The disastrous
      effects of James Watt's tenure as Secretary of the
      Interior appear again and again, each time embellished in
      a different context. Zaslowsky emphasizes the emergence
      of the preservation of ecosystems as a wilderness value
      in the chapter on wilderness areas but also notes its
      growing importance in the management of all public lands.
      Contrary to received opinion, he sees wilderness itself
      as multiple use encompassing protection of watersheds,
      wildlife,   recreational   opportunities,   educational,
      scientific and aesthetic values, and species diversity.
      This is an excellent background source by a writer in
      complete command of his material.

PERSONAL, POLITICAL AND SCIENTIFIC STATEMENT

BIOLOGY AND ECOLOGY

80   *Bioscience* 38.7 (July/August 1988).
      A *Bioscience* special issue devoted to the recently
      organized field of conservation biology.   What is new
      about conservation biology and what separates it from
      traditional biology, on the one hand, and environmental-
      ism, on the other, is its mixture of systematics, field
      study and advocacy.   The articles included here all
      demonstrate dedication to both formal scientific investi-
      gation and to the preservation of biological diversity and
      natural ecosystems.   Conservation biology is not value
      neutral. Conservation biologists see their discipline in
      terms of its contribution to maintaining some areas of the
      planet in a wild state so that both genetic and phenotypic
      variation of species can continue indefinitely.

81   Bates, Marston.   *The Forest and the Sea: A Look at the
          Ecology of Nature and the Economy of Man*. New York:
          Random House, 1960.
      An informal introduction to ecological concepts enlivened
      by references to the author's own experiences studying
      mosquitoes and other insect population in Africa and the
      Amazon.   Marston Bates dedicates this book to his zoology
      students at the University of Michigan whose questionings
      led him to attempt a work that would challenge anthropo-
      centric thinking about the natural world.   The question
      "What good is it?" so often asked of other species assumes
      a "cozy universe in which everything had a purpose for
      man."   The only meaningful question is "What is its role
      in the economy of nature?" Bate's discussion includes
      man's place in nature, and he confronts the issues
      surrounding human growth through technology and sheer
      numbers into a geological force.   "The trend of human
      modification of the biological community is toward
      simplification" and a dismal planet even in our own terms.
      Ethical, aesthetic and utilitarian considerations all
      advocate maintaining wildness and the diversity of life.

His text represents a relatively early attempt to popular-
ize the changing values associated with ecology.

82   Carson, Rachel. *Silent Spring.*   Boston, MA: Houghton
         Mifflin, 1962.
The book that convinced America to begin the process of
turning away from the use of the dangerous chemicals
developed during World War II to control insect pests.
The author reveals and documents the appalling facts about
the effects of DDT, dieldrin, and other poisonous sub-
stances on soil, water, wildlife, domestic animals, and,
potentially, on human beings.   She also exposes the
venality and the willful ignorance of chemical manufacturers and government agencies in encouraging their indiscriminate use.   This, in itself, was a major task, but Rachel
Carson does more.   She educates her readers to the
philosophic   source   of   environmental   disasters.   The
arrogant "control of nature" rationale of the technicians
and   bureaucrats   ignores   fundamental   biological   and
ecological principles in relying on the primitive notion
that nature exists for the convenience of man.   A civilization capable of poisoning the earth has only the crudest
understanding of the interrelationships of living things
with their environment.

83   Darling, Frank Fraser. *Wilderness and Plenty.*   Boston,
         MA: Houghton Mifflin, 1970.
A collection of the Reith Lectures delivered by Frank
Fraser Darling for the BBC in 1969.   As Paul Brooks points
out in his introduction, this is an "ideal book for those
who wish to learn, within a brief compass, what the
environmental crisis is all about."   Darling weaves his
theme of the association of wilderness with plenty
throughout the course of the lectures.   The aesthetically
and ecologically impoverished way of life brought about
by a vicious spiraling of population, technology and
narrow economic values is repeatedly contrasted with the
ecological wealth of the wilderness.   Wilderness is the
"great natural buffer," an "active agent in maintaining
a habitable world."   Darling is not optimistic about the
future without cultural and ethical change that would
inculcate respect for nature and recognition of the
membership of the human species in the earth community.
"The exclusion of man from the hierarchy of nature, so
common in the past and even in our own time, is to put him
in the position of a bourgeois rentier, living off an
economy but having no responsibility for it."

84   Dasmann, Raymond. *A Different Kind of Country.*   New
         York: Macmillan, 1968.
An ecologist's plea for the preservation of diversity.
The author's concern goes beyond wilderness preservation
to include maintaining and creating diverse suburban and
city environments.   We are rapidly diminishing the
opportunity to live any but a controlled, restricted,
homogeneous kind of life in a world where diversity of
living things and variety of habitats have all but
disappeared.   We must learn to plan against progress or
face the terrifying prospect of what happens when there

is no more "outside."   Dasmann demonstrates how the
insights of ecology can be applied to solving problems of
growth and giving direction to conservation.

85   Ehrenfeld, David.   *Conserving Life on Earth.*   New York:
          Oxford University Press, 1972.
     A combination of descriptive ecological science, examina-
     tion of causes of environmental degradation in human
     activity and culture, and discussion of the value of wild
     species and communities to humankind.   The author, a
     biologist, applies the concepts of ecology to the purposes
     of conservation and argues that only when ecological
     realities are recognized can conservation efforts be
     successful. The second chapter provides a clear introduc-
     tion to ecological terminology for the layperson.
     Ehrenfeld is not hopeful that the plea for conserving the
     diversity of natural communities will be heeded in time
     to ward off further disasters. What he fears most is the
     passing of the rich and varied natural world, which has
     heretofore been the stimulus for intellectual activity,
     artistic creation and simple daily pleasures.

86   Ehrlich, Anne, and Paul Ehrlich.   *Extinction: The Causes
          and Consequences of the Disappearance of Species.*
          New York: Random House, 1981.
     An introduction to the problem of species extinction
     famous for the author's use of the comparison between
     destroying species to popping rivets from an airplane--at
     some point the plane and, by analogy, the ecosystem
     crashes.   A combination of an informal style with wide
     ranging scholarship makes this an understandable and
     lively overview not only of the causes and consequences
     of extinction but also of the values--altruistic, economic
     and ecosystemic--that support preservation of wild
     species.

87   Krutch, Joseph Wood.   *The Great Chain of Life.*   Boston,
          MA: Houghton Mifflin, 1956.
     Polished personal essays organized to develop and elabo-
     rate two themes: the close emotional and evolutionary
     relationship between man and other forms of life and
     conscious awareness as evidence of the existence of value
     in the natural selection process. For Krutch watching the
     dying and sudden stillness of a primitive, multi-celled
     creature reveals unequivocally the significance of the
     discontinuity between the living and the nonliving.
     Everything that lives has characteristics that the
     nonliving never have. Observations of a pet salamander's
     modest feats of learning, on the other hand, inspire a
     meditation on what we mean by higher animals and a
     speculation that awareness and emotion have value beyond
     their contribution to survival.   The best of these essays
     are superbly written with terse epigrams flowing effort-
     lessly from the author's pen.   This is especially so of
     "Reverence for Life," where Krutch sums up much about
     human greed and economic organization in a few sentences:
     "The hard heart is more economically productive than the
     tender one.   The sporting instinct pays off.   Reverence
     for life does not."

88   Leopold, Aldo. "A Biotic View of the Land." *Journal of Forestry* 37.9 (September 1939): 727-730.
A pioneering presentation of the idea of land as the foundation of a biotic pyramid composed of an up circuit of food chains culminating in the large carnivores and a down circuit of death and decay returning nutrients to the soil.   In this short essay, delivered as part of a Symposium on Land Use sponsored jointly by the Society of American Foresters and the Ecological Society, Leopold identifies major misunderstandings of the biosphere and of the enormity of the effects of human interventions on its functioning that persist even today.   The contention still heard that the extinction of species as the result of human action is no different from the extinctions that occur normally in the evolutionary process is countered here with clear explanations and examples.   Leopold's use of the terms "violence" and "violent" repeatedly through-out the text adds rhetorical power to what is basically scientific exposition.

89   Lovelock, J. E.   *Gaia: A New Look at Life on Earth.* Oxford: Oxford University Press, 1987.
A lightly revised edition of the author's 1979 book tracing his search for evidence of the hypothesis that the "Earth's living matter, air, oceans, and land surface form a complex system which can be seen as a single organism" able to keep the planet a fit place for life.   J. E. Lovelock calls this self-regulating system Gaia after the ancient Greek earth goddess.   Like the parts and subsys-tems of an individual organism, the living species that make up Gaia maintain through their biological processes a chemical composition of the atmosphere that could not be predicted from the atmospheric compositions of our planetary neighbors.   The Gaian hypothesis is an alterna-tive to the pessimistic view of nature as an alien force to be subdued.   But Gaia is not invulnerable.   Though the biosphere has been able to adjust to major disruptions in the evolutionary past, the speed of mankind's brutal disturbance of natural ecosystems could undermine Gaia's cybernetic functions.   Protecting critical environments, Gaia's vital organs, in their natural state from the effects of increasing human population will be necessary if we are to avoid the "final choice of permanent enslave-ment on the prison hulk of spaceship earth, or gigadeath to enable the survivor's to restore a Gaian world." Lovelock hopes that our instinctive sense of the world's beauty when we live in a balanced relationship with other forms of life may enable us to accept a new role as their partner.

90   Martin, Vance, ed.   *For the Conservation of Earth.* Proceedings of the 4th World Wilderness Congress, Denver, Colorado.   11th-18th September 1987. Golden, CO: Fulcrum, 1988.
A collection of papers and speeches from the 4th World Wilderness Congress organized in four thematic sections: The Global Challenge; National Case Studies; Economics, Development and Environment; and Culture, Man and Environ-ment.   Presenters include environmental professionals,

representatives of conservation organizations, statesmen, economists, representatives of native peoples, and even an artist and a Jungian analyst. The pieces are widely divergent in length and style, but certain ideas and concerns crystallize the tenor of the meeting. Participants repeatedly refer to the importance of wilderness in maintaining the earth's rapidly eroding support systems. There is wide recognition that in an increasingly urbanized world wilderness will not exist unless it is deliberately set aside and protected. Countries must change their ways of doing business. No longer can development projects be proposed and funded without assessing their environmental impact and their sustainability. The issue of the effects of rampant human population growth on wilderness and natural ecosystems is faced squarely by several contributors. Others emphasize the need for education and cultural change if we are to avoid human and ecological catastrophe. The words of native American Jo Agguisho (Oren R. Lyons) of the Onandaga Nation on the necessity for living in accordance with the natural law of the earth eloquently summarize the concerns of the congress: "We are blessed and we prosper when we live by the law. It is dark, terrible and merciless if we transgress the law."

91 Myers, Norman. *A Wealth of Wild Species: Storehouse for Human Welfare*. Boulder, CO: Westview, 1983.
A detailed argument for the economic value to human beings in their everyday lives of the continued existence of a diversity of wild species. The author, although conscious himself of a sense of identification with wildlife, learned from responses to his earlier book *The Sinking Ark* that many people are more interested in the instrumental value of wild species. As a result, he set out to discover specific examples of this kind of value. His wide research is condensed here into a fascinating story of ways wild species contribute to maintaining the variety, nutritional value and health of food crops and to developing effective medicines and substitutes for non-renewable resources. Myers makes clear that genetic engineering does not take place in a vacuum; the earth's genetic stocks form the raw materials for devising useful combinations and serve as inspiration for the creativity of biotechnologists. Continuing to force species to extinction at the current rate will in the end severely handicap our ability to improve the quality of human life in the future. This is an accessible source of solid information on a topic that is often treated only in general terms. It includes a lengthy bibliography.

92 Norse, Elliott A. *Ancient Forests of the Pacific Northwest*. Washington, DC: Island Press, 1990.
A scientific examination of the values of and threats to the old-growth forests of the Pacific Northwest written in language that "makes sense for thoughtful laypeople, yet has the rigor required by professionals." Ecologist Elliott Norse interweaves his own broad knowledge with the results of the most recent research to define Westside forest ecosystems, analyze their biological values, and

assess threats to their viability from timber operations
and global warming. Ancient forest systems create soils,
minimize erosion, store carbon, cleanse the air and
maintain biological diversity. Forestry as it is currently practiced degrades these ecosystem services. It is
"not renewable resource management. It is not even
farming. It is mining." And it will ultimately destroy
the biodiversity on which all forest values, utilitarian
and aesthetic, rest. Norse recommends that opposing sides
in the logging controversy work together to create a
knowledge-based sustainable yield forestry that will put
an end to destruction, fragmentation and simplification
of forest ecosystems and ensure their preservation for a
world that needs both wilderness and wood.

93   Norton, G. Bryan, ed. *The Preservation of Species: The
      Value of Biological Diversity.* Princeton, NJ:
      Princeton University Press, 1986.
Insightful analyses and interpretations of scientific data
on the extinction of species for the purpose of addressing
and solving conservation policy issues. This interdisciplinary collection was planned under the auspices of the
Center for Philosophy and Public Policy at the University
of Maryland to include the work of original thinkers in
the disciplines of biological science, law, resource
management, and philosophy. Major sections cover definition of the problem, values and obligations, and management considerations. In his epilogue the editor summarizes the areas of consensus among the contributors. All
recognize that the coming wave of extinctions is unprecedented in numbers, in the compressed length of time over
which they are projected to take place, and in the
progressive simplification of the ecosystem that will
result. All agree that the most persuasive utilitarian
argument for the preservation of species in their wild
habitats is the possibility of a serious ecosystem
disruption due to cascading extinctions. The simplified
biotic world that would remain would not be a pleasant
place for human beings. Among the nonutilitarian reasons
is the value to human beings of an aesthetically beautiful
and interesting world.

94   Soulé, Michael, ed. *Conservation Biology: The Science
      of Scarcity and Diversity.* Sunderland, MA: Sinauer,
      1986.
A collection of writings by conservation biologists
designed to provide a synthesis of the discipline, to
encourage communication among all sectors of the
conservation community, and to inspire a sense of
excitement and purpose in students and professionals. The
work is organized in subject sections covering the fitness
and minimum viability of populations, the implications of
diversity and rarity for conservation, the effects of
fragmented habitats, the importance of community stability
and structure, the management of habitats, and the
problems of interacting with the political and social
world. For the student of wilderness values the
introductory chapter by the editor and the closing essay
by Arne Naess are the most significant. Both authors deal

with the need for scientists to speak out in the "real world" on behalf of the intrinsic value of the nonhuman world. As Soulé writes, the "planetary tragedy is also a personal tragedy" to scientists who must witness "so much termination." They need the courage to share their values with a general public that looks to them for leadership. Otherwise, in spite of the genuine concern of the majority, habitual consumerist preferences will prevail.

95 Western, David, and Mary C. Pearl, eds. *Conservation for the Twenty-first Century.* New York: Oxford University Press, 1989.
Selected papers from Conservation 2100: A Fairfield Osborn Symposium, a conference of biologists and wildlife experts held in New York in 1986. The book is organized around the "common goal of how to conserve nature in the twenty-first century" with a special emphasis on wildlife. The papers are grouped in four sections. The first gives an overview of the impact of increasing human population on the resources and environment of the coming century. The second and third sections, titled "Conservation Biology" and "Conservation Management," bring together technical articles that analyze the extinction problem or discuss how scientific knowledge can be applied to assure the success of conservation efforts. The fourth section, containing papers examining the "bases for an environmental ethic in American society," is the most relevant for wilderness values. Here Eugene Hargrove summarizes the history of environmental ethics and its place in philosophy; Holmes Rolston, III builds a case for human conscience as the evolutionary development that defines our biological obligation to conserve the diversity of life; and Bryan Norton argues that human values of diversity, contrast and solitude provide sufficient reason for preserving wild nature. The collection concludes with a precis of material presented at the conference by Michael Soulé, an agenda for the future and a helpful bibliography.

96 Wilson, E. O., ed. *Biodiversity.* Washington, DC: National Academy Press, 1988.
Papers from the National Forum on Biodiversity held in Washington, on September 21-24, 1986, under the auspices of the National Academy of Science and the Smithsonian Institution. According to the editor, the forum coincided with increasing interest among scientists and portions of the public in biodiversity brought on by the accumulation of data on species extinction and new awareness of the linkage between conservation and economic success in developing nations. The brief, well-documented papers are arranged in thematic sections that access current conditions, both globally and in selected threatened habitats, or explain recent scientific or management techniques for monitoring and preserving species. Throughout the collection contributors stress the central role of habitat destruction in species extinction, the potential commodity value of diverse species as food or pharmaceuticals, and our obligation to keep the earth habitable for future

generations.   Part 5, "The Value of Biodiversity," and
Part 12, "Ways of Seeing the Biosphere," are especially
relevant to the study of wilderness values.   The papers
in the former explain or debate the appropriateness of
using economic concepts to evaluate biodiversity and the
worth of species.   Those in Part 12 explore the emotional
and spiritual aspects of membership in the biosphere.

97   Wilson, Edward O.   *Biophilia*.   Cambridge, MA: Harvard
          University Press, 1984.
Essays by sociobiologist Edward Wilson exploring the
notion of biophilia, the innate tendency in human beings
to focus on life and lifelike processes.   The author uses
his fieldwork in Surinam and elsewhere to introduce his
view that humans have been shaped by evolution, not only
physically but mentally, to prefer, even require for their
normal development, an environment that includes other
living things.   The human brain is peculiarly adapted to
receive experiences of nature that include the savannah-
like landscapes of our distant ancestors.   People react
more fully and quickly to organisms than to machines.   Our
art is overwhelmingly made up of motifs from the natural
world.   The controlled, mechanized environment devoid of
other life that we are busy creating for ourselves is the
result of a tragic ignorance of human nature.   The
collection also includes an essay entitled "The Conservation Ethic," which suggests the phylogenetic continuity
of life as a reason to preserve nonhuman species.

98   Wilson, Edward O.   "Million-Year Histories: Species
          Diversification As an Ethical Goal." *Wilderness*
          48.165 (Summer 1984): 12-17.
A reminder by the well-known sociobiologist that mankind
evolved in interaction with the nonhuman natural world
and, so, is uniquely attuned to its diversity of sight and
sound.   Wilson ponders the idea that we may need wilderness not only because it produces useful and potentially
useful plants and animals but as the home of the human
spirit.   The natural world is intrinsically interesting
to human beings in a way that artificial environments are
not.

CRITIQUES OF WESTERN CULTURE

99   Berry, Thomas.   *The Dream of Earth*.   San Francisco, CA:
          Sierra Club Books, 1990.
A passionate book that interprets the present as a time
of radical transformation and calls upon contemporaries
to create a new historical vision to guide the process of
renewing our relationship with the natural world.   Western
culture with its dream of a millennium in which humans are
exempt from the consequences of being an integral part of
the community of earth is creating a wasteland of pollution and extinction rather than its promised "wonder-
world."   Technological cunning employed in the service of
a pitifully narrow conception of human nature and human
good is depriving us of the wild earth as the location of
"splendid and intimate modes of divine presence."   Yet,

as a Catholic monk and a historian of culture, Thomas
Berry recognizes Western traditions as necessary contributors towards a redeemed human presence in the world.   The
Western idea of historical change is itself a positive
force in the task of creating a new civilization.   The
creation theology of Christianity, long eclipsed by the
theology of personal salvation, carries the seeds of
religious revival. Most important, science, the "Yoga of
the West," has given us a knowledge of the nature and
functioning of earth and life that is adequate to preserve
and protect them.

100   Ehrenfeld, David.   *The Arrogance of Humanism.*   New
         York: Oxford University Press, 1978.
An attempt to expose some false and arrogant assumptions
that underlie the conceptual framework of modern humanism.
The core of humanism from which lesser assumptions derive
is "a supreme faith in human reason--its ability to
confront and solve the many problems humans face, its
ability to rearrange both the world of nature and the
affairs of men and women so that human life will prosper."
David Ehrenfeld uses the technique of end-product analysis
to show that, when many of the triumphs of humanism are
seen in any but a short-term, immediate context, they
prove destructive of both human potential and the well-
being of the planet. In the chapter called "The Conservation Dilemma" he reveals the inadequacy of resource
conservation arguments to ensure the preservation of wild
species and wilderness: rational humanists will not save
natural communities that lack practical value as commodi-
ties. Rather than using exaggerated estimates of economic
value in piecemeal attempts to save this or that species,
we need a philosophy or religion that demands that natural
communities and species be protected "*because they exist
and because this existence is itself but the present
expression of a continuing historical process of immense
antiquity and majesty.*"

101   Haines, John.   *Living Off the Country: Essays on Poetry
         and Place.*   Ann Arbor, MI: University of Michigan
         Press, 1981.
Speculations on the meaning of wilderness interspersed
among essays on the state of modern poetry and the role
of place in forming the poetic imagination.   For John
Haines "wilderness is not just that state of mountains,
rivers, and forest, but a condition that lives in the
universe," one that is the source of aliveness and
awareness in human beings as well as in other animals.
Still, for humans, culture is as necessary as wilderness;
dissatisfaction and decay occur only when culture has
succeeded in walling out wilderness.   Contemporary
American civilization offers a narrow, derivative existence, where we live off the surface of things, blind to
the monstrous effects we have on the earth in our pursuit
of wealth and security.   The wilderness experience of
solitude, silence and simplicity reconfirms authentic
existence.   In a crowded world, where for most people it
is no longer possible to enter the wilderness except
sporadically, poetry rooted in place can awaken us to the

immensity of nature and to our true selves. And finally, if the human enterprise comes at last to nothing, the "wilderness is here, ready to spring back into place like the branches we push aside as we walk through the forest."

102 Livingston, John A.   *The Fallacy of Wildlife Conservation*.   Toronto, Canada: McClelland and Stewart, 1981.
A stripping away of the rhetoric from arguments for wildlife preservation to reveal their powerlessness in the face of the scarcely conscious belief structure of Western civilization. At the root of this structure is the fear of our inevitable mortality. Civilization hides from us the reality of our vulnerable biological being with conceptual givens that separate humanity from the rest of the biosphere and render us unable to perceive the world except through the lenses of our dominance over the nonhuman and our prime place as the apex and end of evolution. Logical arguments, whether in the form of resource conservation or ethics, are ultimately impotent because our mythology assures that the short-term needs of the technomachine will always prevail. A solution, if there is one, will come from individuals learning one by one, in spite of social conditioning, to open themselves to the experience of belonging in and being part of wild nature. Livingston writes frankly, forcefully, and often humorously, having eschewed flattery and optimism. He appears to be a pessimist by any definition; yet there is something invigorating and hopeful in his refusal to play the game with the old stacked deck.

103 Lopez, Barry Holstun.   *Of Wolves and Men*.   New York: Scribner's, 1978.
An exploration of the ways humans have imagined and treated wolves since ancient times. In the first few chapters Barry Lopez summarizes what we actually know about wolves from observation and field biology; then, in the major portion of the work, he examines the wolf of human mythology and imagination--the noble hunter emulated by Eskimos and Pawnees, the close companion of the Norse god Odin on the battlefield, and, most important for Western history, the evil beast going about the Devil's business. Lopez draws a parallel between the American settlers' fear and destruction of the wilderness and their implacable hatred of the wolf. The wolf is the symbol of the wilderness, of all that is wild and primitive and inimical to civilized life. The near hysterical pogroms that have come close to exterminating wolves go beyond rational predator control and even beyond casual cruelty to become the "violent expression of a terrible assumption: that men have the right to kill other creatures not for what they do but for what they may do." We have yet to learn to live peacefully in a universe that contains other beings with their separate realities.

104 McKibben, Bill.   *The End of Nature*.   New York: Random House, 1989.
An assessment of the significance of global warming and a lament for the passing of nature that it heralds.

Although Bill McKibben carefully reviews the scientific predictions concerning the effects of increased levels of carbon dioxide in the earth's atmosphere, his main concern is the change that vagrant, artificially induced climate and weather will bring about in the idea of nature. Heretofore, in spite of local pollution and destruction, nature has remained independent of humankind. Pristine places have been the source of solace and security for humans sickened by social ills. We could enter the wilderness and be assured of permanence and harmony and of belonging to something larger than ourselves. The voice of the God of Job came out of a whirlwind to praise His nonhuman creation; isolated in our man-dominated world our religion will come to us through the religious broadcasting networks. We could refuse to "compound original sin with terminal sin" by reducing our numbers and living more humble lives. The more likely outcome, should we avoid disaster in the interim, is the genetically engineered and macro-managed planet touted by the technological optimists. McKibben may or may not be correct in his account of the seriousness of global warming. His enduring contribution is an uncompromising presentation of the present as a crossroads in the relationship of humans to the natural world.

105   Snyder, Gary. *Good Wild Sacred*. Madley, Hereford, UK: Four Seasons Press, 1984.
An explication of the concepts good, wild and sacred as they apply to the land. Gary Synder contrasts the world picture of the Australian aborigines and the Japanese Ainu people with that of European culture and the modern industrial state. In these hunter-gatherer societies wilderness--the unmanipulated and unmanaged natural environment--is identical with good land supportive of human life. All the environment partakes of the sacred although special places concentrate spiritual power. In contrast, the modern world totally separates these concepts. Sacredness, except in the eyes of wilderness advocates, is confined to temple precincts; and even good, productive land is routinely degraded by exploitative monoculture farming. Synder foresees a possible cultural revolution heralded by the deep ecology and Earth First! movements as developing a new synthesis where "wild, sacred, and good will be one again and the same, again."

106   Stone, Christopher D. *Should Trees Have Standing? Legal Rights for Natural Objects*. Los Altos, CA: Kaufmann, 1974.
A proposal to award natural objects the right to sue in the courts on their own behalf and to be beneficiaries of legal actions. The author compares natural objects, on the one hand, to corporations, which are considered legal entities although they are not persons, and, on the other hand, to incompetent humans, who may be represented in legal actions by guardians. Through their guardians, natural objects would become jural entities capable of representing the significant but fragmented and diffuse damages to wilderness users and supporters when government agencies or corporations propose development that would

destroy wilderness values. Stone's essay, which appeared initially in the *Southern California Law Review*, is quoted by Justice William O. Douglas in his dissenting opinion from the Supreme Court decision of April 19, 1972, a decision which refused the Sierra Club standing to represent Mineral King Valley in opposing a Forest Service action allowing Disney Enterprises to construct a ski resort in the area. Both the majority and the dissenting opinions are appended to Stone's text.   The book also contains a perspicacious foreword by Garrett Hardin.

107   Tobias, Michael, ed.   *Deep Ecology*.   San Diego, CA: Avant Books, 1984.
An eclectic collection of writings held together by their relevance to the rich mixture of science, sociology, politics, poetry and philosophy that make up the deep ecology program.   The pieces are arranged in three sections: "Hard Facts" analyzes history and trends; "Heartland" combines poetry and mythology; and "Awareness and Reason" examines the grounds for environmental ethics. Most of the articulate supporters of deep ecology are represented here, as well as writers, such as, Herman Daly and Garrett Hardin, who are noted for their ideas on contemporary economic and resource problems.   George Sessions contributes a piece on the recent history of attitudes toward nature that calls needed attention to the role of the beatniks and hippies of the 1950s and 1960s in preparing the ground for an ecological consciousness. Dolores La Chapelle and Alan Grapard pay tribute to eastern insights into wilderness values by examining, respectively, Taoism and Japanese religions and mythologies.   James Dickey's poem, "For the Last Wolverine," captures in harsh imagery and abrupt rhythms the helplessness and rage for revenge he feels at the destruction of wild species.   As a whole, this is a worthwhile and provocative anthology.

EDUCATION

108   Bowen, James.   "Science, Education and the Environment: Ecocentrism as the New Paradigm."   *Education and Society* 6.1-2 (1988): 3-15.
An argument for the necessity of reforming science education to reflect a holistic model of the relationship of humankind to nature.   The author demonstrates how our dominant reductionist science has contributed to a technology and way of life that are unsustainable yet is unable, because of its anthropocentric model of man as outside of nature, to generate solutions to our emerging ecocrises. The contrasting ecocentric paradigm is gaining importance as a theoretical corrective to reductionism, but it has yet to influence education. Bowen suggests the work of the early progressives in education, especially, Pestalozzi, as a starting point for a re-examination of educational philosophy.

109   Cohen, Michael J. *How Nature Works: Regenerating Kinship with Planet Earth*.   Portland, OR: Stillpoint, 1988.

A text for learning about nature through seeking awareness of ourselves as part of Gaia, the living earth. Michael Cohen shows how Audubon Expedition Institute courses employ all the senses--radiation, feeling, chemical, mental and spiritual--as well as symbols and intellect to teach knowledge of the natural world and our involvement in it. Each chapter concludes with a study guide that encourages students to learn from their own experiences and to explore their own feelings and imaginings. Many of the exercises and activities in the study guides can be adapted for use in classroom situations. The goals of the AEI program are listed in the final chapter; and an appendix gives directions for obtaining detailed information about courses, workshops and registration procedures.

110   Cohen, Michael J.   *Our Classroom is Wild America: Trailside Education in Action--Encounters with Self, Society, and Nature in America's First Ecology Expedition School.*   Jericho, NY: Exposition Press, 1974.
The story of the Trailside School operated during the 1970s by Michael J. Cohen and his wife Diana as an alternative to one year of the traditional high school curriculum. Although the school is headquartered on a 200 acre wildlife sanctuary in Vermont, students spend extended periods of time camping out in "varying communities, ecosystems, microenvironments, and habitats throughout the United States." The school attempts to educate people to a more joyful and rewarding way of life by integrating and blending five principles: (1) human beings are members of and dependent on an ecological community: (2) direct contact with teachers and other students in a real-life environment encourages personal growth; (3) small group living provides feedback for individual development: (4) meaningful learning occurs when students are totally involved in the learning situation; and (5) freedom to reach individual decisions and assess their consequences is necessary for developing responsibility.   Throughout the book the author gives specific content to these principles by recreating dialogues with students.

111   Cohen, Michael J.   *Prejudice Against Nature: a Guidebook for Liberation of Self and Planet.*   Freeport, ME: Cobblesmith, 1983.
A book of insights and advice derived from the author's many years of experience as director of the National Audubon Society Expedition Institute, an accredited educational travel community for college and high school students.   The programs and teaching methods of the AEI are meant to "reverse the destructive trends of our culture." Michael Cohen sees America's unquestioning acceptance of the exploitation and destruction of natural systems as the result of a prejudice against nature instilled in children at an early age. This prejudice is so pervasive that experts in the field of human behavior do not even recognize, much less study, our sustaining relationships with the natural world.   An outline of content and objectives for courses of study that investi-

gate aspects of prejudice against nature is included among the appendices.

112 Nash, Roderick, and others.   "Course Descriptions in Environmental Studies Part Three: Special Topics." *Environmental Review* 8.4 (1984): 348-374.
A collection of syllabuses for various college courses that focus on humankind's relationship to wilderness and the natural world. Although they are offered by departments in the sciences and the humanities or in career programs, these courses are interdisciplinary in that they include a mixture of readings--scientific, legal, philosophical and literary. Their topical organizations and challenging assignments make them a resource for researching the ways in which changing wilderness values can be incorporated into postsecondary education.

113 Sagoff, Mark. "The Philosopher As Teacher: On Teaching Environmental Ethics." *Metaphilosophy* 11 (1980): 307-325.
An exposition of how one teacher solved the problem of adapting ethics, usually taught in courses organized around personal morality, to environmental problems that require public action and collective responsibility. Mark Sagoff argues that "environmental problems are moral problems primarily in the sense that they are political and legal problems." They result from our failure as a society to recognize the difference between our preferences as individuals and the general will. What we need is not so much a change in personal ethics as ecologically sensitive social norms that will prevent the institutionalization of individual or group preferences in order to protect the values of the community. Sagoff's students provided him with an illustration of how this might work in their reactions to the case of the Sierra Club's opposition to a Disney sponsored ski resort in the Mineral King wilderness. Although the students would individually prefer a resort, as members of the community they opposed compromising the wilderness with recreational facilities. Teachers looking for ways to involve college students in environmental debate will find helpful suggestions in this article.

114 Yambert, Paul A., and Carolyn F. Donow. "Are We Ready for Ecological Commandments?" *The Journal of Environmental Education* 17.4 (1986): 13-16.
An overview of problems environmental professionals have encountered in teaching environmental ethics, together with a proposed solution in the form of an environmental code that can be used to guide students in developing an environmentally ethical lifestyle. The authors, a professor and researcher of forestry, identify the ideal person from an ecologist's point of view as one who has reached the highest stage of moral development, characterized in Kohlberg's developmental model as centering on the unity of the cosmos. However, since few people reach this stage, a concerted effort is needed to change the basic assumptions of the average student about man's relationship to nature. Yambert and Donow suggest that this can

best be accomplished by teaching general rules of ecologically sound behavior through discussions of particular problems and issues.

SIERRA CLUB WILDERNESS CONFERENCES

115 Brower, David, ed. *The Meaning of Wilderness to Science.* Proceedings of the 6th Wilderness Conference. San Francisco, 1959. San Francisco, CA: Sierra Club, 1960.
A collection of addresses by biological scientists focused on the value of wilderness to science. The basic insight that emerges from these proceedings is that wilderness areas provide scientists a natural laboratory for studying the interrelationships of plants and animals within an environment and for understanding the complexity of ecological and evolutionary mechanisms. This idea is clearly enunciated in Frank Fraser Darling's "Wilderness, Science and Human Ecology." Darling outlines the devastation and ecosystem simplification brought about by excessive grazing in the Highlands of Scotland and that area's consequent uselessness for learning about complex conversion cycles and energy flows. In the baldest terms, then, wilderness is necessary to provide "study areas of pristine conditions." The collection, as a whole, is an excellent source for examining one type of instrumental value attributed to wilderness.

116 Brower, David, ed. *Wilderness: America's Living Heritage.* Proceedings of the 7th Wilderness Conference. San Francisco, 1961. San Francisco, CA: Sierra Club, 1961.
A collection of speeches and papers organized around the theme of the value of wilderness to American culture. Participants discuss the influence of the wilderness on the national character and on American art. More politically oriented discussions focus on the outlook in Congress for the passage of S.174, the latest wilderness bill, which had been recently introduced by Senator Clinton P. Anderson. Congressman John P. Saylor, who introduced the first wilderness bill in 1956, assesses the situation in Washington. Other influential contributors include photographic artist Ansel Adams, Sigurd Olson, Joseph Wood Krutch, Secretary of the Interior Stewart L. Udall, and Justice William O. Douglas.

117 Brower, David, ed. *Wildlands in Our Civilization.* San Francisco, CA: Sierra Club, 1964.
A selection of papers from the first five wilderness conferences sponsored by the Sierra Club between 1949 and 1957. Most notable for their concern with wilderness values are A. Starker Leopold's "Wilderness and Culture" and Bruce M. Kilgore's "Wilderness and the Self-Interest of Man." Starker Leopold echoes his father's conviction that the only motivating force for wilderness preservation that will persist in the long run is moral, the belief that we owe the earth's wild communities protection from development. Kilgore argues that we need to preserve

wilderness because we may, in the headlong rush of development, destroy organisms and processes that may be necessary for improving and maintaining human life. The anthology concludes with summaries of the remarks made at each of the five conferences.

118   Kilgore, Bruce, ed.   *Wilderness in a Changing World.*
           Proceedings of the 9th Wilderness Conference, San
           Francisco, 1965.   San Francisco, CA: Sierra Club,
           1966.
Addresses and discussions from the first Wilderness Conference held after the Wilderness Act of 1964 was passed. Although most contributors touch briefly on the Act as evidence for a changed public attitude toward wilderness, the emphasis of these writings is on the problems that remain--pollution, reclamation, and, especially, population growth.   Representatives from administrative agencies reflect on the impact of the Wilderness Act with muted enthusiasm and considerable worry about implementation.   In contrast, the section entitled "What Wilderness Means to Man" contains some of the most thoughtful and passionate writing to appear in the proceedings of any of the conferences. Anthropologist Ashley Montagu's "Wilderness and Humanity" deserves special mention for its perceptive consideration of what the devastation of the natural world is likely to do to a humanity no longer able to see itself as part of a natural whole but condemned instead to exist in an environment composed solely of lifeless manmade things.

119   Leydet, Francois, ed.   *Tomorrow's Wilderness*: Proceedings
           of the 8th Wilderness Conference.   San Francisco,
           1963. San Francisco, CA: Sierra Club, 1963.
Papers and discussions from the last Wilderness Conference to be held prior to the enactment of the Wilderness Act in 1964.   The ideas and remarks of the participants reflect a mixture of guarded optimism over the near-term legislative situation and fear for the long-term future of wilderness.   Stewart Udall, then Secretary of the Interior, concentrates this concern in a concise parallel-ism: "As a culture develops, wilderness is the last resource to acquire value.   As a culture feels the pressure of population, wilderness will be the first to be consumed." Fairfield Osborn goes into greater detail, enumerating the many hopeful signs in recent years but setting in the balance against them the detrimental effects of relentless population growth.   Wilderness values form the subject of papers by James P. Gilligan and Paul Brooks.   Gilligan reports the findings of the University of California Wildlands Research Study Center; and Brooks surveys our evolving sense of the importance of wilderness. His conclusion, at least for the immediate future is a happy one: "We have the votes and we intend to be heard."

120   McCloskey, Maxine, ed.   *Wilderness: The Edge of
           Knowledge.*   Proceedings of the 11th Wilderness
           Conference.   San Francisco, 1969.   San Francisco,
           CA: Sierra Club, 1969.

A collection of speeches, papers, and discussions influenced by the new ecology movement that was beginning to
gather strength at the time. The book is organized in two
broad areas: the impact of human activities on the health
of wilderness ecosystems and Alaska as the last frontier.
A major emphasis of the first section is the importance
of wildlife to wilderness ecology. The extensive treatment of Alaska reflects the fact that in 1969 the Alaskan
lands settlement had yet to take place, while work on the
oil pipeline had already begun. As is true of all the
Wilderness Conferences, these proceedings bring together
the ideas and opinions of legislators and conservationists, including such major figures as Senator Henry M.
Jackson and Roderick Nash. The appendices include the
texts of the Endangered Species Act of 1969 and of the
Regulations on Public Outdoor Recreation Use of the Bureau
of Land Management Lands.

121 McCloskey, Maxine, and James P. Gilligan, eds.
Wilderness and the Quality of Life. Proceedings of
the 10th Wilderness Conference. San Francisco,
1967. San Francisco, CA: Sierra Club, 1969.
A rich collection of expressions of passionate concern for
the wilderness, factual reviews of the implementation of
the Wilderness Act, and documents from the Forest Service
and the National Park Service on identifying and managing
wilderness areas. There is a special section on the
contribution of wilderness to American life and another
on desert wildernesses. Speakers include cabinet
secretaries and heads of government agencies, as well as
conservation leaders, scientists and scholars. The theme
of the quality of life runs through the different speeches
and discussions and reveals the many values, from economic
to spiritual, that wilderness has for Americans.

122 Schwarz, William, ed. *Voices of the Wilderness.* New
York: Ballantine, 1969.
An anthology of papers and speeches from the Sierra Club
Wilderness Conferences held during the 1960's. The
selections are arranged in three sections. The second
section, entitled "The Values that Wilderness Preservation
Seeks to Preserve," includes pieces by such spokesmen for
the wilderness as William O. Douglas, Sigurd Olson and
Joseph Wood Krutch, as well as several articles by
scientists on the value of wilderness to science. This
is an excellent source for gaining an understanding of the
modern wilderness movement.

WILDERNESS AND WILDLIFE MANAGEMENT

123 Allen, Durward L. *Our Wildlife Legacy.* Revised
edition. New York: Funk & Wagnalls, 1962.
The revised edition of a book originally published in 1954
as a guide to aid sportsmen's and other groups with an
interest in developing strong state and national wildlife
programs in forming a conservation philosophy based on
ecologically sound wildlife and game management. Durward
Allen sees wildlife as a public asset or resource, part

of the American standard of living, that is being threatened by policies reflecting political expediency and
biological ignorance. He shows how programs once thought
to protect and increase desirable game species--stocking
with hatchery or farm raised animals, misconceived bag
limits and predator control--ignore ecological principles
and, so, fail in accomplishing their stated aim. The most
important factor in preserving wildlife is preserving
habitat. A new conservation ethic demands that we
exercise the measure of social self-control necessary to
set aside wilderness areas where complete animal-plant
complexes can continue to exist and afford an opportunity
for Americans "to view their country as it once was."
Durward's book has special historical interest in that it
combines in a somewhat unstable compound elements of
emerging ecological concern and respect for the wild with
an older emphasis on game production. Words like "har-
vest" and "crop" are used throughout in reference to
animal populations.

124 Baker, Ron. *The American Hunting Myth.* New York:
       Vantage Press, 1985.
An angry indictment of modern sport hunting and wildlife
management practices inspired by the author's experiences
as a wildlife refuge landowner in New York. Ron Baker
confronts the arguments that hunters as a group are
proponents of wilderness preservation and that hunting is
necessary to control the population of game species by
exposing the ecologically destructive effects of the
prohunting bias of state wildlife laws and practices.
Government conservation agencies derive much of their
income from hunting licenses and taxes on firearms; so it
is hardly surprising that they adopt policies that result
in persecution of predators, neglect of nongame species,
and production of unnaturally large populations of favored
game animals, especially the large ungulates. Buttressing
this systematic catering to the demands of special
interests is a professional education program that
perpetuates the view that animals have no value except as
resources for human consumption. Baker gives some useful
specific suggestions for reform through gradually shifting
funding of fish and game bureaus to general revenues and
monies obtained from nonconsuming users of public wildlands. His hope is that the development of a biocentric
cultural ethic will finally make recreational killing
unacceptable.

125 Devall, Bill, and George Sessions. "The Development of
       Natural Resources and the Integrity of Nature."
       Environmental Ethics 7.3 (Winter 1984): 292-322.
A wide-ranging article combining philosophy with personal
conviction and recommendations for land management. The
authors' chief purpose is to demonstrate how current and
traditional resource conservation and development management programs are rooted in scarcely articulated philosophical assumptions about the world and man's place in
it--humans are the central figures, and all the rest is
a resource for humans. In contrast, the norms of the deep
ecology world view are self realization, in the sense of

moving from the isolation of the ego toward beginning to
identify with human and nonhuman others, and biocentric
equality, in principle, of all creatures. Conservation
management based on deep ecology would stress minimum
impact on ecosystems.

126 Hendee, John C., George H. Stankey, and Robert C. Lucas.
Wilderness Management. Forest Service Miscellaneous
Publication No. 1365. A 1.38:1365. Washington, DC:
U. S. Forest Service, 1978.
A carefully organized exposition of the role of wilderness
management in implementing the National Wilderness
Preservation System. Although the discussion is directed
to environmental professionals and includes chapters on
planning by objectives and determining carrying capacity,
the language is straightforward and the content of
interest to a broader audience. The authors define
wilderness management as "essentially the management of
human use and influence to preserve naturalness and
solitude." There was a transitory moment in conservation
history when recreational use of wildlands could exist
without management. Today, in an age of rapidly expanding
wilderness use, only wise management can protect the
values--the wilderness experience, mental and moral
renewal, and ecosystem integrity--the Wilderness Act
sought to preserve. The authors recommend a management
approach they term biocentric as contrasted anthropocentric. Their biocentric philosophy recognizes the legiti-
macy of recreational use but emphasizes the goals of
protecting the experience of solitude and permitting
natural ecological processes to operate freely. In the
final analysis, human enjoyment depends on taking wilderness on its own terms.

127 Line, Les, ed. *What We Save Now: An Audubon Primer of
Defense*. Boston, MA: Houghton Mifflin, 1973.
A collection of some of the best nature writing from
Audubon magazine by authors who are "proud to be called
preservationists." They include such well-known naturalists as Joseph Wood Krutch, Edwin Way Teale, and Sigurd
F. Olson and cover topics as diverse as duck baiting, the
ecology of Florida swamps, and the intrusion of the "awful
ORVs" on the wilderness experience. There is much good
science and good writing here. Although nearly twenty
years have passed since most of these pieces were written,
few of the issues that outraged these environmentalists
in the 1970s have been adequately resolved. The conflicts
between preservationists and those the editor calls "the
despoilers" over endangered species, off-road vehicles,
pesticides and pollution highlighted in this volume
provide a cultural context for our current environmental
problems.

128 Mitchell, John G. *The Hunt*. New York: Knopf, 1980.
An exploration of why hunters hunt from the point of view
of a writer who is uncertain in his own mind about the
morality of hunting. Mitchell partakes as an observer and
questioner in three very different kinds of hunting, what
he terms the banal, the exotic, and the sublime. His

first hunt, opening day of deer season in Michigan, involves hordes of orange-coated hunters, many of them city folk with little knowledge or love of the land, descending on the woods and looking for a shooting place as near the road as possible. For the second he goes to the privately owned game ranches of Texas that guarantee an exotic ungulate for the right price. The third is hunting at its best, a week long elk hunt camping out in the Gallatin Divide of Montana with companions who care as much about being alone in wild country as they do about shooting a trophy. Mitchell finds among these and other skillful, dedicated hunters strong supporters of wilderness and wildlife. On his last trip he joins them in a deer hunt.

129   Sax, Joseph L. *Mountains without Handrails: Reflections on the National Parks*. Ann Arbor, MI: University of Michigan Press, 1980.
An appeal for goal clarification in management of the national parks and other public wildlands for recreation. The author argues convincingly that these areas cannot provide a setting for all kinds of recreation and tourism. Their unique recreational value is to offer the opportunity for humans leading increasingly controlled lives in a technological and mechanized environment to experience themselves and the natural world directly without the mediation of machines or set agendas. The parks should provide settings for people of differing degrees of physical strength, wilderness experience and outdoor skills to partake in contemplative, not consumptive, recreation. Those forms of recreation that are most consumptive, especially, power cycling and snowmobiling, have no place in the wilderness.

130   Schoenfeld, Clarence A., and John C. Hendee. *Wildlife Management in Wilderness*. Pacific Grove, CA: Boxwood Press, 1980.
A manual for "translating wilderness philosophy into practical tactics" addressed to personnel in federal and state agencies charged with management of wildlife in Congressionally designated wilderness areas, to students preparing for wildlife careers, and to members of outdoor and conservation groups. The central theme is that within wilderness areas wildlife and resource management should be governed by commitment to the wilderness ideal of maintaining natural ecosystems as wholes with their indigenous wildlife populations in naturally occurring numbers. Wilderness wildlife managers need to understand that in these areas wilderness values take precedence over recreation or other uses of public lands. Wild animals play a major part in how we value wilderness; and wilderness, in turn, functions to protect endangered species and species with highly specialized habitat needs. The authors are especially concerned that the needs of wilderness dependent species, for example, wolves and grizzlies, for freedom from human contact not be sacrificed to multiple-use criteria or traditional game management practices. The book concludes with clear

guidelines and proposed objectives for successful wilderness management.

131 Simon, David, ed.   *Our Common Lands: Defending the National Parks*. Washington, DC: Island Press, 1988. Legal essays by lawyers and law professors exploring the strengths and weaknesses of the laws available to protect the National Park System from the effects of activities that would impair its mission to conserve the scenery, natural and historical objects, and wildlife of the parks so as to leave them "unimpaired for the enjoyment of future generations." The collection opens with an assessment of the threats to park values and the inadequacy of the Park Service's attempts to deal effectively with them. The bulk of the essays examine the ways in which particular laws can contribute to park preservation. Part II concentrates on general legal authorities, such as, the National Park Service Organic Act of 1916 and the Federal Land Policy and Management Act. Part III covers legislation protecting specific park values, the Clean Air Act and the Endangered Species Act, for example, while Part IV addresses additional laws that apply to energy, mining and dam-building. Summing up this comprehensive coverage, the editor draws the conclusion that the Park Service and concerned citizens have substantial existing legal and moral authority to protect the parks from diverse threats. The problem has been in the failure adequately to exercise and test this authority. Special features of this volume include a case study of Glacier National Park by Joseph L. Sax and Robert B. Keiter and an appended "Basic Primer on Legal Source Materials for Nonlawyers."

WILDERNESS DEFENDERS

132 "The Wilderness Society Platform." *The Living Wilderness* 2.2 (November 1936): 15. The eleven point statement of the principles of the Wilderness Society. Most important historically is the statement's emphasis on wilderness as a natural mental resource, which bears the same value for mankind's psychological needs as physical resources, such as, timber or water, bear for their material needs. Wilderness is a need, not a "luxury or plaything".

133 De Voto, Bernard.   *The Easy Chair*.   New York: Houghton Mifflin, 1955. A collection of Bernard De Voto's essays originally published in the Easy Chair feature of *Harper's*. The section entitled "Treatise on a Function of Journalism" reprints selected pieces from the series of articles he published between 1947 and 1954 exposing the attempts spearheaded by wealthy western stockmen and sheep growers to discredit the Forest Service and slip through Congress bills that would return public lands to the states for later resale to private interests. The fascinating story of how De Voto discovered their intentions and planned his opposing maneuvers is told in the Notes section. Had these bills become law they would have effectively

dismantled the program of natural resources conservation begun by Teddy Roosevelt; but their success depended on stealth and secrecy, acting before the public became aware of what was actually being proposed beneath the rhetoric of individual freedom and state's rights. De Voto's journalism made the "landgrab" national news, thus, stymieing the earliest and most comprehensive of the bills. He, then, kept up the pressure over a period of years with these hard-hitting essays grounded in his scholar's knowledge of western history.

134   Douglas, William O.   "Wilderness and Human Rights." *Voices for the Wilderness* . Ed.   William Schwarz. New York: Ballantine, 1969: 109-121.
Justice Douglas' conviction that human beings need wilderness, as well as civilization, in order to develop wholeness of being is shared by many conservationists. What is unique in this essay is his insistence that the experience of a healthy natural world is a human right. The right to hear the "song of the whippoorwills at dawn in a forest where the wilderness bowl is unbroken" should be guaranteed and protected as are our civil rights. This essay also appears in the Proceedings of the 7th Biennial Wilderness Conference entitled *America's Living Heritage* published in   1961.

135   Douglas, William O.   *A Wilderness Bill of Rights.* Boston, MA: Little, Brown, 1965.
An outline of Justice Douglas' concept of wilderness as a right of the American people.   Douglas sees wilderness as providing Americans with these important values: a place to relive American history; temporary escape from the pressures of urban life; unparalleled opportunity for wonder and aesthetic experience; education in the interdependence of all life.   A Wilderness Bill of Rights would assure for the minority who love wildness and for posterity the same protection that the Bill of Rights gives to other minorities.   Its ingredients would include restraints on multiple use of federal lands, legislative wilderness areas, control of fencing on public lands, and protection against dams, highways and motorized recreation.   Broader protections against air and water pollution would also be needed.   An office of conservation would be created to counteract the industry orientation of existing agencies responsible for administering public lands.

136   Frome, Michael.   *Conscience of a Conservationist: Selected Essays.*   Knoxville, TN: University of Tennessee Press, 1989.
A collection of conservation essays published between 1969 and 1980.   Approximately half the articles deal with preserving wild lands in Southern Appalachia; the rest tackle conservation issues throughout the nation.   Michael Frome is a plain-spoken, hard-hitting journalist whose refusal to submit to censorship cost him his job as conservation editor for *Field & Stream* magazine in 1974. His writings document the dedication of citizen activists in the struggle to gain recognition for protection of our

national wilderness heritage and the rights of future
generations from representatives of myopic and uninterest-
ed federal agencies. They also offer an opportunity to
observe the process of changing wilderness values as it
was reflected in the inability of the Forest Service and
the National Park Service to comprehend the involvement
of ordinary people in implementing the Wilderness Act.
By the late 1960s the narrow economic values which seemed
to dominate both agencies no longer made sense to millions
of Americans concerned about endangered species and
ecosystems in an increasingly urban world.

137   Haines, John, et al. *Minus 31 and Wind Blowing: 9*
         Refections about Living on the Land. Anchorage, AK:
         Alaska Pacific University Press, 1980.
A collection of essays adapted from conference and
workshop lectures given in the 1970s under the sponsorship
of the Alaska Humanities Forum. Most deal with attitudes
towards the Alaskan land or explore the place of wilderness in Alaska's future. John Haines recalls the events
and emotions of three days in coldest winter alone tending
his traps in the wilderness. Historian Thomas LeDuc
analyzes the history of American natural resources policy
as political irresponsibility, materialism and the idea
that wilderness should be "tamed and subjected to man's
uses." Margaret Murie contrasts the values of people who
come to Alaska to get something or to get away from
something, the boomers and the rip-off types and the
escapers, with those who love Alaska for what it is, its
wildness and space, and "change themselves in order to
meet its conditions." Poet Gary Synder closes his essay
and the book with a revised pledge of allegiance that
celebrates Turtle Island, the diverse American ecosystem
and all the beings who dwell within it "with joyful
interpenetration for all."

138   Henberg, Martin. "Wilderness as Playground." *Environmental*
         Ethics 6 (1984): 251-263.
An exploration of the value of wilderness as the necessary
environment for a unique kind of play closely related to
American history. Martin Henberg recognizes the importance of other wilderness values, aesthetic appreciation
and preservation of biodiversity, for example; but he
focuses on wilderness as a place to playfully mimic the
experiences of early explorers and mountain men and, thus,
to participate in the history and folklore of the westward
expansion. This kind of cultural play requires the
remoteness of wilderness to maintain an illusion that is
easily shattered by motorized recreation and other forms
of play that are not truly wilderness dependent.

139   Krutch, Joseph Wood. "Human Life in the Context of
         Nature." *Voices for the Wilderness.* Ed. William
         Schwarz. New York: Ballantine, 1969: 161-168.
A thoughtful essay organized around two common sayings:
"No man is an island" and "This is one world." The author
extends their meaning to include all living things and the
natural world as a whole. In the modern world the
acceptance of these truths is necessary because we now

have the power, or nearly so, to eliminate all other
complex life and to bring about an earth which will
contain nothing but man and that which serves his
immediate needs.   The essay ends with an expression of
fear for the future: "...if we continue to act as though
men were mere machines, they may actually in the end,
become something very like machines."   This essay
originally appeared in the Proceedings of the 7th
Wilderness Conference entitled *Wilderness: America's
Living Heritage.*

140   Leopold, Aldo. "Why the Wilderness Society."   *The
        Living Wilderness* 1.1 (September 1935): 6.
        A statement of the philosophical importance of the
        founding of the Wilderness Society that appeared in the
        first issue.   The Wilderness Society is a "disclaimer of
        the biotic arrogance of 'homo americanus,'" "one of the
        focal points of a new attitude," and evidence of an
        "intelligent humility toward man's place in nature."

141   Leopold, Aldo.   "Wilderness Values."   *The Living
        Wilderness* 7.7 (March 1942): 24-25.
        A wartime expression of two major themes in wilderness
        preservation and in Leopold's mature thought: wilderness
        as a necessary contrast to life in society and wilderness
        as an organic process.   Both these values are degraded by
        motorized recreation and by wildlife management that
        controls natural predators to provide an excess of game
        for hunters.

142   Marshall, Robert.   "The Problem of Wilderness."   *The
        Scientific Monthly* February 1930: 141-148.
        An early public call for the establishment of wilderness
        areas, as distinct from national parks or forests.
        Marshall's logic, clarity and occasional satiric wit make
        this an effective statement of the value of wilderness for
        human beings.   The article is structured in the manner of
        a debate.   First, the benefits of wilderness are present-
        ed; next, the drawbacks are given in rebuttal; finally,
        the two sides are evaluated, and a plea for action
        formulated.   The section on benefits details the role of
        wilderness in promoting physical health, psychological
        health, and aesthetic experience.   This brief article by
        one of the founders of the Wilderness Society is an
        important document in the history of the wilderness
        movement.

143   Marshall, Robert.   "The Universe of the Wilderness Is
        Vanishing."   *Nature Magazine* 29 (April 1937): 235-
        240.
        An examination of the unique value and fragility of
        wilderness and a plea that these be fully considered in
        making decisions that involve conflicts of genuine values.
        Wilderness is the source of an overwhelming aesthetic
        experience of the unity and antiquity of the natural world
        that is, nevertheless, easily diminished by the artifici-
        ality and fragmentation that result from roads and
        automobiles.   The greatest good for the greatest number
        does not mean that every acre of public land has to be

devoted to the maximum of resource production or mass recreational use. Quality as well as quantity must be taken into account. Wilderness elicits a depth of experience that may be worth an "infinite number of ordinary experiences." Robert Marshall believes the United States can "afford to sacrifice almost any other value for the sake of retaining something of the primitive."

144   Murie, Olaus. "Wild Country As a National Asset." *The Living Wilderness* 18.45 (Summer 1953).
A special issue of *The Living Wilderness* devoted to the lectures given by Olaus Murie while he held the Isaac Hillman Lectureship in the Social Sciences at Pacific University in April 1953. The overall theme of the three lectures ("God Bless America-and Let's Save Some of It," "Wild Country Round the World," and "Beauty and the Dollar Sign") is the need to save wild landscape and wild creatures "not for what materials they furnish, but for the help they might be to us by simply remaining mountain, desert, forest, and river...." Murie admits that he cannot fully explain the human yearning for wild country and the spiritual contentment it brings; yet he fears for our culture and our essential humanity should it no longer be available to future generations. His own writing is infused with that humanity as he relates the experiences not only of writers like Thoreau but of the ordinary people he has met in the wilderness or at hearings and meetings.

145   Nash, Roderick, ed. *Grand Canyon of the Living Colorado.* San Francisco, CA: Sierra Club, 1979.
A book of photographs and essays brought together both to celebrate the success of conservationists in defeating Bureau of Reclamation Central Arizona Water Project plans to build dams at Marble and Bridge Canyons within the Grand Canyon and to generate support for completing and extending Grand Canyon National Park. Contributions include evocations of rafting or walking the canyon wilderness by backpacker Colin Fletcher, photographer Ernest Braun and historian Roderick Nash. Though the tone and style of their writing differs, all three record experiencing a sense of belonging to an immense order of time as they enter the canyon and adopt the rhythms of the wilderness, what Nash calls "cosmic contentedness." They contrast this with the narrow, short-term exploitative values that would degrade a timeless landscape to spur material growth in desert cities for a few decades. In a separate piece Nash traces the history of the controversy over building dams in the canyon and argues for enlarged national park status to assure adequate protection against future attempts to compromise its integrity. The collection also reprints the Sierra Club newspaper advertisements that were instrumental in marshalling the letter writing campaign that convinced the 1968 Congress to refuse authorization for any dams within the Grand Canyon.

146 Nash, Roderick. "Wilderness: To Be or Not to Be?"
        Nature and Human Nature. Ed. William R. Burch, Jr.
        Yale University School of Forestry Bulletin No. 90.
        New Haven, CT: Yale University, 1976.
     A perceptive enumeration of wilderness values presented
     as one of the International Championship lectures at Yale
     University during the 1974-1975 academic year. Roderick
     Nash offers this paper in response to the questioning of
     the importance of wilderness preservation by environmen-
     talists concerned with the survival problems of population
     growth and pollution. Nash is especially successful in
     expressing the meaning of "wilderness as an aid to
     developing environmental responsibility." Wilderness has
     educational and symbolic value for "building stable, long-
     term harmony between man and environment." It teaches
     humility and respect for the land through allowing us to
     rediscover natural processes and our dependence on them
     and to experience ourselves as part of a community which
     does not exist simply to fulfill man's needs. As a
     symbol, it proclaims our commitment to restraint. By
     establishing wilderness preserves, Americans affirm that
     we "put other considerations before growth."

147 Nash, Roderick Frazier. "Why Wilderness?"
        Conservation of the Earth. Proceedings of the 4th
        World Wilderness Conference. Denver, Colorado.
        11th-18th September 1987. Ed. Vance G. Martin.
        Golden, CO: Fulcrum, 1988.
     An articulation of wilderness values that draws a clear
     distinction between good and bad arguments for defending
     wilderness. Roderick Nash is convincing in his contention
     that, in a time when wild places are hard pressed by
     demands of expanding civilization, the importance of
     wilderness preservation needs to be conveyed by a sound
     wilderness philosophy. He identifies three fallacious
     arguments for preserving wilderness--beautiful scenery,
     recreation that is not wilderness dependent and economic
     advantage. The enumeration of good arguments that follows
     emphasizes ecological, spiritual and heritage values and
     concludes with the nonanthropocentric argument that
     nonhuman life and ecosystems have intrinsic value. Nash
     sees wilderness preservation as a truly radical act
     because it challenges the growth ethic. Arguments that
     appeal to that ethic are ultimately counterproductive.

148 Olson, Sigurd F. *Reflections from the North Country.* New
        York: Knopf, 1976.
     A collection of essays summing up the personal beliefs
     Sigurd Olson arrived at over a lifetime of wilderness
     experiences in the Quetico-Superior canoe country, Alaska,
     and throughout the United States and Canada. Weaving
     philosophic discussion with examples from his own experi-
     ence, he conveys his conviction that wilderness is
     necessary because it encourages us, even forces us, to
     realize who we really are--beings shaped by our evolution
     through millennia to find wholeness through being aware
     and in tune "with waters and rocks, with vistas and
     horizons, with constellations and the infinity of space
     and time." In wilderness we moderns have the opportunity

to live for brief periods the timeless gathering and nomadic life of our ancestors and to carry back to civilization the peace and stillness this brings.

149   Schaefer, Paul.   *Defending the Wilderness: The Adirondack Writings of Paul Schaefer.* Syracuse, NY: Syracuse University Press, 1989.
A Collection of essays written over a period of fifty years to defend the "forever wild" covenant of the New York State Constitution and protect the Adirondack wilderness against recurring threats to its integrity from dam-building projects and motorized recreation.   These articles, many of which appeared in local conservation magazines and newsletters, alerted the citizens of the state to the consequences of various bills and amendments for the wildlife, rivers and old-growth white pine forests of the park preserve.   Paul Schaefer was an early member of the Wilderness Society and a friend of both Bob Marshall and Howard Zahniser.   His book is especially valuable for his recollections of their conversations together along the trails of this wild country.

150   Stegner, Wallace.   *The American West as Living Space.* Ann Arbor, MI: University Press of Michigan, 1987.
An overview of the gestalt of the American West by a writer for whom both the West as region and the West as state of mind are native habitat.   The essays in this slim volume were originally delivered as three William C. Cook Lectures at the Law School of the University of Michigan in 1986.   In them Wallace Stegner elaborates on the recurring themes that elucidate the meaning of the West: aridity, reclamation, public lands, open space and the myth of the independent, self-reliant cowboy.   Historically, the public lands and the agencies that administer them have been the battle ground of conflicting wilderness values.   Do they exist to provide cheap grass, coal, timber, or predator control for favored stockmen and lumbering and mining companies?   Should they serve the recreational needs of a much larger public?   Or do they have a higher duty to maintain the health and beauty of the land?   Behind the massive dams and the destructive development projects that disfigure the western landscape stands the blindly arrogant doctrine that the "destiny of man is to possess the whole earth."   Stegner suggests a different doctrine: to learn to live with this arid country on its own terms.

151   Stegner, Wallace.   "Living on Our Principle." *Wilderness* 48 (Summer 1985): 15-21.
A passionate assessment of our progress in adopting a land ethic since the publication of Leopold's *A Sand County Almanac* in 1949.   On the side of improved relationships with the land the author counts legislation, such as, the Wilderness Act and the Endangered Species Act, as well as an astounding growth in the power of environmental organizations.   On the negative side, Stegner sees little improvement in the development of an environmental conscience governing our use of the land.   The easily educable have already been reached, and the rest are as

willing as they were in Leopold's day to sacrifice the
long-term health of ecosystems to exploitative greed or
personal pleasure.

152   Stegner, Wallace.   *The Sound of Mountain Water.*   New
         York: Dutton, 1980.
A collection of the author's nature writings on the
American West and his interpretations of western literature written during the period from 1946 to 1968.   The
essays on Glen Canyon and Lake Powell should be used with
those of Edward Abbey to gain a picture of the beauty of
that lost wilderness.   The most eloquent piece in the
collection is the author's 1960 letter in connection with
the report of the Outdoor Recreation Resources Review
Commission.   Stegner speaks of the importance of wilderness in American life. Wilderness, for we Americans, who
were shaped by its existence even as we fought to subdue
it, is the "geography of hope."

153   Stegner, Wallace, ed.   *This Is Dinosaur: Echo Park
         Country and Its Magic Rivers.*   New York: Knopf,
         1955.
The book that was instrumental in arousing public opinion
against building a hydroelectric dam and flooding the
canyons of Dinosaur National Monument in the mid-1950s.
At the time Dinosaur was a comparatively little known part
of the National Park System.   This book, with its photographs and chapters devoted to the archaeological,
geological, historical, wildlife and recreation values of
the monument, was written to show the American people what
they would be giving up if they allowed their wilderness
park to be replaced by a reservoir.   The tone is not
polemical; nevertheless, some powerful arguments for
wilderness preservation emerge in the essays.   Novelist
Wallace Stegner uses his rhetorical skill to praise the
special form of human mark on the environment which is the
"deliberate refusal to make any mark at all."   In the
concluding chapter Alfred A. Knopf emphasizes the serious
consequences for conservation of authorizing a dam-
building project that compromises the inviolability of the
national parks.

154   Zahniser, Howard.   "The Need for Wilderness Areas."
         National Parks Magazine October-December 1955:
         161-166+.
An article based on a paper presented at the National
Citizen's Planning Conference on Parks and Open Spaces for
the American People held May 24, 1955.   Zahniser's paper
marked the initial volley in the nearly decade long effort
to gain legislative protection for wilderness areas.   As
Executive Director of the Wilderness Society, Zahniser
reiterates the theme of wilderness as a psychological and
spiritual need rather than simply an amenity that was
sounded in the platform of the society at its founding.
That theme is enriched by the author's emphasis on the
educational value of wilderness.   Modern science has
demonstrated that human beings are members of the community of life, dependent on the whole for their existence;
yet the technological sophistication of modern life tends

to hide that dependence and deceives us into imagining that we can sustain ourselves apart from the whole. Wilderness opens our eyes to reality.

PHILOSOPHY

155 Armstrong-Buck,   Susan.   "Whitehead's   Metaphysical
        System as a Foundation for Environmental Ethics."
        Environmental Ethics 8 (1986): 241-260.
     An attempt to demonstrate that Whitehead's philosophy
     provides an explanation for our intuitive recognition of
     value in nature and an adequate metaphysical foundation
     for the development of widespread cultural support for the
     preservation of wilderness.   After discussing the main
     tenets of Whitehead's thought, the author compares them
     with the theories of Peter Singer and Tom Regan, which
     locate value in sentient beings, and with Leopold's land
     ethic and recent ecological interpretations of Spinoza,
     which attribute value to wholes.   She is particularly
     cogent in showing how Whitehead's concept of creative
     actual occasions and their organization into societies
     could support the Leopoldian view of the value of ecosys-
     tems.

156 Attfield, Robin.   *The Ethics of Environmental Concern.*
        New York: Columbia University Press, 1983.
     A defense of the Judaeo-Christian concept of stewardship
     as sufficient ethical grounds for granting moral consider-
     ableness to both future generations and nonhuman living
     beings.   The author argues convincingly that the concept
     of man as having dominance over but no obligations toward
     the natural world represents an unjustifiable selectivity
     in interpreting the scriptures and a failure to adequately
     examine   Christian   and   Jewish   historical   practices.
     Attfield's own philosophical position is a variety of
     consequentialism which equates ethical action with the
     maximization over time of intrinsic value residing in
     living beings.   This is a difficult book that presumes
     familiarity with a host of writers in the field.   It is
     included   here   because   it   represents   an   alternative
     rationale for valuing wilderness.

157 Birch, Charles, and John B. Cobb, Jr.   *From the Cell to
        the Community.*   Cambridge, England: Cambridge
        University Press, 1981.

A collaborative effort on the part of biologist Charles
Birch and theologian John B. Cobb, Jr., to introduce a new
cultural paradigm based on the ecological model for the
purpose of liberating individuals and institutions from
the deadening effects of faulty belief systems.   Two
cultural models stand in the way of respect for nature:
the mechanistic model of an outmoded science inherited
from Descartes and other 17th century thinkers and the
religious model of human beings as rational souls separated from nature.   Although these models were rendered
obsolete by Darwin's discoveries in the 19th century, they
remain operative because of fear, linguistic convention,
and convenience.   The ecological model proposed by the
authors is based on Whitehead's process philosophy and
posits events rather than substance as the fundamental
reality.   In substance thinking all relations are exter-
nal.   Relations in events are internal; they define the
event.   Human beings are societies of events that include
nonhuman others both in their pasts and in their presents.
All living things possess intrinsic value because they are
manifestations of the cosmic power that brings creative
novelty out of chaos and entropy.   Birch and Cobb, Jr.,
call this power Life and equate it with God.   Much of the
book details how acceptance of this new understanding of
the world could bring about a sustainable society with
justice for all.

158   Birch, Thomas H.   "The Incarceration of Wildness:
         Wilderness Areas as Prisons." *Environmental Ethics*
         12 (1990): 3-26.
An explication and condemnation of the cultural meaning
of setting aside legally designated wilderness areas.
Thomas Birch argues that the value of wilderness areas
lies in their otherness; they are the purest expressions
of that wildness at the heart of existence which refuses
to be defined and limited by civilization.   In establish-
ing wilderness reserves, the "imperium" attempts to
appropriate wild nature by placing it within the domain
of human law.   Rather than changing our fundamental
relationship with the land by welcoming the spontaneous
manifestations of wildness in woodlots, suburbs and
cities, we lock it up in circumscribed sacred space to be
visited occasionally for recreation.   Even so, wilderness
reserves cannot be completely managed.   As "holes" or
"cracks" in the "fabric of domination and self-deception
that fuels our mainstream culture," they maintain their
subversive potential.

159   Brennan, Andrew. *Thinking about Nature*.   Athens, GA:
         University Press of Georgia, 1988.
An argument for respecting nature and wild things based
on what scientific ecology reveals about reality.   The
author sets out to show that an environmental ethics that
recognizes wholes or ecosystems and nonliving entities,
such as, rivers and mountains, as morally considerable can
be justified without accepting the metaphysical ecology
of the deep ecologists.   Scientific ecology demonstrates
that wholes have properties and powers that are more than
the sum of their parts.   It also shows that individuals

and species, in addition to their inherited characteristics as autonomous wild creatures, possess functional
qualities within an ecosystem. They are what they are in
part because of their context. Western moral theories
have ignored the importance of context in determining what
is ethical. That we humans are part of nature existing
within a complex system for sustaining life is sufficient
grounds for respecting the natural world and taking the
welfare of nonhuman beings and the global system into
consideration in making political and moral decisions.
Brennan shares with Christopher Stone the view that
morality is pluralistic and that it is unnecessary and
probably impossible to construct a single framework for
arbitrating moral judgments.

160   Callicott, J. Baird.   "The Conceptual Foundations of
        the Land Ethic."   *Companion to *A Sand County*
        Almanac: Interpretive and Critical Essays*.   Ed. J.
        Baird Callicott.   Madison, WI: University of
        Wisconsin, 1987: 186-217.
A closely reasoned explanation of the divergence of
Leopold's land ethic from mainstream modern ethical
theory, which bases the criteria for ethical consideration
on the extension of egoism to include all rational beings
or, in its more expansive form, all sentient beings, but
which cannot admit the claims of wholes, such as, biotic
communities, to moral status.   Leopold, in contrast,
derives his ethic from the dissenting arguments of Hume
and Darwin, who postulated inherent sympathies and
feelings as the source of moral codes. Leopold envisions
an ethical obligation to the ecological whole that can
place restrictions on the right of human beings to pursue
their perceived good at the expense of the health of the
biotic community. In his discussion, Callicott effectively counters the contention that an operational land ethic
would require draconian or fascistic measures in its
execution.

161   Callicott, J. Baird.   "Hume's Is/Ought Dichotomy and
        the Relation of Ecology to Leopold's Land Ethic."
        Environmental Ethics 4 (1982): 163-172.
An intellectually neat answer to the mechanical invocation
of Hume's strictures against deriving ought from is by
critics objecting to environmental ethical systems that
locate value in nonhumans or ecosystems.   Callicott uses
Hume's own concept of the existence of altruistic feeling
toward kin and society in all normal human beings to show
that evolutionary and ecological knowledge, which relates
humans to all other creatures in a community of life,
imposes the obligation to act ethically in our dealings
with them.

162   Callicott, J. Baird.   *In Defense of the Land Ethic:
        Essays in Environmental Philosophy*.   Albany, NY:
        State University of New York Press, 1989.
A collection bringing together Callicott's articles and
essays originally published in journals and anthologies.
The author has resisted the temptation to revise his
earlier work for this volume in order to make it possible

for the reader to follow his thought as it changes and develops over a decade. The pieces are clustered in five sections: (1) Animal Liberation and Environmental Ethics; (2) A Holistic Environmental Ethic; (3) A Non-Anthropocentric Value Theory for Environmental Ethics; (4) American Indian Environmental Ethics; and (5) Environmental Education, Nature Aesthetics, and E. T. (extraterrestrial life). Many of these writings are annotated separately in this bibliography, but reading them as they are organized here will facilitate both understanding and appreciating his work.

163 Callicott, J. Baird. "Intrinsic Value, Quantum Theory, and Environmental Ethics." *Environmental Ethics* 7 (1985): 257-275.
A three-part intellectual adventure exploring progressively more encompassing rationales for finding value in the natural world. In the first section the author clarifies his earlier subjectivist arguments conjoining Hume's conception of natural sentiment, the genetic basis of ethics elucidated in sociobiology, and the kinship of life revealed by evolution and ecology. This is an adequate foundation for an environmental ethic; but in the second section Callicott finds it wanting because it assumes a sharp distinction between subject and object that the developments of 20th century science have made no longer tenable. All qualities, including values, can now be seen as secondary qualities existing potentially but actualized in the interaction between conventional subject and conventional object. The third section pushes beyond conservative interpretations of the new physics to speculate that relations are logically prior to things; and the extended self, therefore, coterminous with the world. If the self has intrinsic value, then, so does the world.

164 Callicott, J. Baird. "'Just the Facts, Ma'am'." *The Environmental Professional* 9 (1987): 279-288.
A clear exposition of the author's major ideas on environmental ethics for an audience outside the discipline of philosophy. Here his emphasis is on the misapplication of the fact/value dichotomy in determining the source of conflicts over wilderness preservation. Callicott argues that, contrary to general opinion, such conflicts and disagreements are not about ultimate values; for ultimate values have been normalized by natural selection. Instead, they arise over proximate values and, so, are amenable to resolution through understanding the relevant facts. Aldo Leopold's about face on predator control serves as a case study for demonstrating the efficacy of knowledge of the facts in changing proximate values. What stands in the way of sound environmental attitudes today is a legacy of obsolete world views that are inadequate to interpret the complexity of rapidly increasing environmental information.

165 Callicott, J. Baird. "The Land Aesthetic." *Companion to A Sand County Almanac: Interpretive and Critical*

Essays. Ed. J. Baird Callicott. Madison, WI:
University of Wisconsin, 1987: 157-161.
A thought-provoking essay that illuminates the extent to
which Leopold saw beauty and aesthetic values as impor-
tant--perhaps, more important than ethics--to the preservation of wilderness. The author argues that Leopold
created a new aesthetic for the appreciation of nature
based not on the artificial beauty of the picturesque but
on a joining of direct sensory perception, involving all
the senses, with an awareness of ecological and evolutionary relationships. This is a seminal article for the
study of the aesthetics of nature.

166 Callicott, J. Baird. "The Metaphysical Implications of
Ecology." *Environmental Ethics* 8 (1986): 301-316.
An incisive explanation of how the world picture drawn by
the science of ecology and the new physics effectively
dissolves problems of attributing value to the natural
world created by the still dominant world view of modern
classical science. Ecology, like physics, is a field or
matrix science, where individual organisms appear as
temporary formations in the flow of energy; organisms are
implicated in and imply the whole. The distinction
between self and other is blurred; and consciousness, no
less than physical characteristics, is seen to have
developed in relation to the evolution of ever more
elaborate and diversified environments. Given this kind
of interwoven world, human concern and love for wild
things and dismay at environmental degradation is the
normal and appropriate response.

167 Callicott, J. Baird. "Non-Anthropocentric Value Theory
and Environmental Ethics." *American Philosophical
Quarterly* 21 (1984): 299-309.
The most complete summary of Callicott's ethical thought
prior to his consideration of the implications of the
field theories of ecology and physics for environmental
ethics. The article is especially useful for the clarity
with which it presents the situation of environmental
philosophy after a decade of development. He addresses
the question of whether this new endeavor is an applied
ethics, similar to bioethics, whose task is merely to
extend conventional anthropocentric ethics to new states
of affairs or whether it is instead an exploration of
alternative moral and metaphysical principles. If the
latter, its task is to develop a non-anthropocentric value
theory. Callicott finds wanting the various theories so
far presented--ethical hedonism extended to animals and
ethical conativism because they apply only to discrete
individuals; theistic axiology because it is inconsistent
with modern science; holistic rationalism, in the mode of
Plato and Leibnitz, because it is compatible with indif-
ference to individual welfare.

168 Carlson, Allen. "Nature and Positive Aesthetics."
Environmental Ethics 6 (1984): 5-34.
Lucid justification of the thesis that the natural world,
where it is undisturbed by human activity, is always
beautiful. The literature of wilderness preservation

frequently refers to the positive aesthetic value of
wilderness. The question is how to explain the modern
consensus that nature, unlike art, cannot be aesthetically
displeasing or ugly. Carlson argues that this is due to
scientific, particularly, ecological, explanations of
nature. The categories science employs to make nature
intelligible are those--such as, order, regularity,
harmony, balance, tension, conflict and resolution--that
are descriptive of aesthetic excellence.

169 Devall, Bill, and George Sessions. *Deep Ecology: Living
     As If Nature Mattered.* Salt Lake City, UT:
     Peregrine Smith, 1985.
A rich mixture of the philosophy, poetry, cultural reform,
and political action that make up the deep ecology
movement. The text is structured so that explanations of
ideas or situations are illustrated by graphically
distinct quotations from poets, philosophers and environmental writers--a formatting technique that enables the
authors to convey the context of the movement visually as
well as verbally. Devall and Sessions contrast deep
ecology with reform environmentalism. The latter they
characterize as attempting to preserve wilderness and
natural ecosystems without departing from the anthropocentric world view and resource management perspective of the
dominant culture. Deep ecology espouses contrasting
ultimate norms: Self-realization, in the sense of achieving awareness of oneself as part of a larger ecological
whole, and biocentric equality. It finds its roots in the
minority Western tradition and taps sources in American
nature writing, the world views of primal cultures, the
insights of Tao and Zen, and the feminist questioning of
the assumptions of the dominant culture. Overall, this
book is a successful summarizing of the various currents
of the deep ecology movement.

170 Elliot, Robert, and Arran Gare, eds. *Environmental
     Philosophy: A Collection of Readings.* University
     Park, PA: Pennsylvania State University Press, 1983.
An anthology of essays designed to stimulate thinking
about the need to reassess humankind's relationship with
the natural environment. The papers in part one examine
the ethical significance of future generations in environmental decision-making. Those in part two deal with
questions about the moral considerableness of nonhuman
animals and ecosystems. Part three consists of essays
that explicitly focus on the viability of the Western
Judaeo-Christian-humanist tradition as a basis for an
environmental ethic. Nevertheless, the majority of
contributors touch on an overall theme: the Western world
picture, which places man outside of and in opposition to
the natural world and restricts the definition of ethical
behavior to the social contract model, is no longer
intellectually tenable. Though it continues to play its
role in justifying the pursuit of unrestrained greed in
exploiting natural resources, it has no support in 20th
century science and fails to account for our everyday
experience of value in nature. Evolution, biology and
ecology all demonstrate that humans are part of the

community of living things and depend on the functioning of the biosphere for their existence. Holmes Rolston, III, speaks for other contributors as well as himself when he points out that humans are endowed with naturally selected capacities to value such things as "the water-fall, the cardinal, the blackberry, the warming sun."

171 Fox, Warwick. "Deep Ecology: A New Philosophy for Our Time." *The Ecologist* 14 (1984): 194-200.
A perceptive article outlining deep ecology as intel-lectual movement and elucidating the major problem in developing a deep ecology environmental ethic. The author clarifies the centrality of the holistic or field view of nature to deep ecology positions, then, draws attention to the penchant of deep ecology philosophers for dividing environmental ethics into two contrasting systems--the resource conservation system, which restricts intrinsic value to human beings, and their own egalitarianism, which accords equal worth to all members of the biosphere. Warwick criticizes this position as not only unworkable in practice but conceptually deficient in that it ignores the contribution of environmental philosophers who ascribe differential intrinsic value to organisms depending on their complexity and capacity for richness of experience. Warwick also provides an unprejudiced account of the relationship between deep ecology and Eastern philosophies and meditative traditions.

172 Frankena, W. K. "Ethics and the Environment." *Ethics and Problems of the 21st Century.* Ed. K. E. Goodpaster and K. M. Sayre. Notre Dame, IN: University of Notre Dame Press, 1979: 3-20.
A review of various types of ethical orientations toward the environment as they are characterized by the answer to the question of what kinds of things have moral relevance. The author identifies eight types: ethical egoism (self); forms of humanism or personalism (human beings); all conscious sentient beings; all living things; all things (includes forms of holism); theisms (God); combinations of types; cooperating with or following nature (nature is moral). Frankena believes that a type three ethics--all conscious sentient beings--is both logically defensible and sufficient for environmental preservation. This relatively brief essay serves as a readable background for more detailed discussions of specific ethical positions.

173 Godfrey-Smith, William. "The Value of Wilderness." *Environmental Ethics* 1 (1979): 309-313.
A succinct article categorizing the instrumental values generally attributed to wilderness by conservationists and hypothesizing that a true environmental ethics is depen-dent on replacing Cartesian dualism with a holistic or total-field view of organisms as nodes in a biotic web. Such a community view of the world can serve as conceptual background for the development of the capacity to empa-thize with what was previously alien and apart. The author outlines the shifts in attitudes which will be necessary before sympathy for the environment becomes

immediate and natural. This relatively early article touches on the limitations of cost-benefit analysis, as well as on the major themes of wilderness philosophy.

174   Goodpaster, Kenneth E. "On Being Morally Considerable." *Journal of Philosophy* 75 (1978): 308-325.
An argument for being alive as both a necessary and sufficient condition for moral considerableness. The author clears the ground for his lucid discussion by separating the notion of moral considerableness from other considerations that often clutter and cloud debate. These include moral significance, which concerns comparative judgments in matters of conflicting moral duties, and moral rights, which may refer to a narrower, legal issue. Kenneth Goodpaster rejects sentience, the capacity to feel pleasure and pain, as a necessary condition for moral considerability because it ignores the scientific evidence that feelings are ancillary to the larger evolutionary capacity of living beings "to maintain, protect and advance their lives," whether they are conscious of doing so or not. Indeed, he speculates that the tendency to limit moral considerability to sentient beings is rooted in the hedonism of a culture that can conceive no good other than pleasure. Goodpaster would seem to locate moral considerability only in individuals; however, he grants that, if the biosphere approximates the behavior of a living thing, it would qualify. This article might profitably be read in conjunction with the work of Paul Taylor.

175   Gray, Elizabeth Dodson. *Green Paradise Lost.* Wellesley, MA: Roundtable Press, 1979.
An exploration of how female consciousness can contribute to creating a less exploitative and alienated way of experiencing and relating to the world. The traditional patriarchal Western world picture takes the form of a hierarchy with man (not woman) at the top a little below the angels and nature, the wild things and the land, at the bottom. Even the Darwinian revolution, which proved this picture false, has been reinterpreted to reserve a special place for mankind as the end and purpose of evolution. This essentially male myth creates blinders that prevent humans from imagining other beings as anything except both strange and inferior. It also produces a false split between mind, higher and male, and body, lower and identified with the female and the bestial. "We do not understand who we are" because myth has separated us from reality. The bodily experiences of women--menstruation, which links them to natural cycles, and pregnancy and childbirth, which relate them to the past and the future--suggest the more humble notion of our human selves as connections in a coordinated, often, symbiotic, web of non-hierarchical diversity. Gray is convinced that, free from the need to bolster our image of superiority and otherness, we could at last be at home on the earth.

176   Hammond, John L. "Wilderness and Heritage Values." *Environmental Ethics* 7 (1985): 165-170.

A defense of cultural heritage value as a legitimate rationale for wilderness preservation in the United States. The author cites Aldo Leopold's contention that wilderness has had a special role in forming the American national character and reviews the literature of psychology to show the importance of assimilating one's roots in the development of mature selfhood. For Americans the experience of wilderness is necessary to achieving a cultural identity.

177 Hargrove, Eugene C.   *Foundations of Environmental Ethics*. Englewood Cliffs, NJ: Prentice-Hall, 1989.
A learned exploration of the roots of environmental ethics in existing Western traditions. The author disputes the contention of certain philosophers that environmental concerns have no place in Western philosophy by demonstrating that the modern environmental outlook is the product of three centuries of changing attitudes toward nature rooted in the aesthetic perceptions associated with the natural history sciences of botany, biology, and geology and embodied in Romantic poetry and American landscape painting. In these traditions wild nature is viewed as intrinsically beautiful and interesting. Hargrove's treatise culminates in a sophisticated ontological argument for environmental ethics that posits natural beauty as fundamentally contingent on physical existence in a way that artistic beauty is not. Existence in nature precedes essence; beauty in nature emerges only as it takes physical form.   Destruction of nature is, thus, inevitably destruction of beauty.   This is a book that should be included in any survey of contemporary positions in environmental philosophy.

178 Hargrove, Eugene C.   "The Value of Environmental Ethics." *The Environmental Professional* 9 (1987): 289-294.
An argument for the importance of environmental ethics to environmental decision-making addressed to doubting professionals. Eugene Hargrove attributes the disappointment many environmental scientists and decision-makers have expressed in the practical usefulness of environmental ethics to a failure to appreciate the role of philosophy as a clarifier of values and principles. Environmental ethics cannot resolve factual uncertainty; what it can do is demonstrate that values are no more subjective than facts and deserve to be taken equally and openly into consideration when decisions are made about environmental affairs. A major effort in "clarifying the ethical issues could be so successful that we might be able in the near future, within the next few decades, to reduce the uncertainty in environmental decisionmaking to factual difficulties alone." The contemporary situation is such that environmental professionals feel themselves compelled to cast their intuitions about the value of, say, wild elephants, in factual terms. A developed environmental ethics would lead to a more honest state of affairs in which values could be openly expressed as reasons that "should influence environmental policy."

179   Hill, Thomas E., Jr.   "Ideals of Human Excellence and
         Preserving Natural Environments."   *Environmental
         Ethics* 5 (1983): 211-244.
      An interesting preliminary presentation of a somewhat
      unusual argument for the preservation of natural environ-
      ments.   The author bids us ask the question about the
      destruction of natural environments for profit or conve-
      nience: "What sort of a person would do that?" He, thus,
      shifts the focus of discussion from arguments about
      intrinsic values or rights to the ideals of human excel-
      lence.   Appreciation of the place of human beings within
      ecosystems and our relationship to other species is seen
      as indicative of the proper humility and self-acceptance
      characteristic of the ideal person.   The reason we feel
      uneasy about willingness to destroy the natural world for
      the benefit of oneself or the group one identifies with
      is that such acts suggest ignorance of the world we live
      in, self-importance, and the denial of our biological
      nature.

180   Kheel, Marti.   "The Liberation of Nature: A Circular
         Affair."   *Environmental Ethics* 7 (1985): 135-149.
      A criticism from the feminist viewpoint of philosophies
      that base arguments for valuing nature solely on rational-
      istic criteria.   No adequate protection for wilderness is
      possible as long as the failure to recognize the impor-
      tance of feeling, emotion and personal experience in moral
      decision-making persists.   Part of the problem stems from
      the tendency of traditional patriarchal ethics to ossify
      into dichotomies and rules.   In contrast, direct involve-
      ment with the natural world permits the fusion of thought
      and emotion into a "unified sensibility," which in its
      highest form is love.   This essay introduces a novel
      approach to valuing wilderness and serves as a welcome
      relief from ethical abstractions.

181   Lehmann,   Scott.   "Do   Wildernesses   Have   Rights?"
         Environmental Ethics 3 (1981): 129-146.
      An examination of the philosophical difficulties found in
      ascribing natural rights or intrinsic value to wilder-
      nesses and other wholes.   Scott Lehmann, a Sierra club
      member as well as a logician, argues that, in spite of the
      attraction of the rights argument to preservationists as
      a device for defending wildernesses, the concept cannot
      be made to apply to entities that are not the subjects of
      experiences.   His summary of his position is succinct:
      only what can be harmed or benefitted can have rights:
      only subjects of experiences can be harmed or benefitted;
      natural objects, except for some animals, are not subjects
      of experiences; therefore, natural objects cannot have
      rights.   He also tackles, though less decisively, posi-
      tions that ascribe value to wild nature as the source of
      intrinsically valuable qualities, such as, beauty.   For
      Lehmann human welfare is an adequate basis for valuing
      wild lands.   The real problems occur in weighing wilder-
      ness against other human values.   His article serves as
      useful background for interpreting the counterarguments
      of ecocentric philosophers.

182 Matthews, Freya. "Conservation and Self-Realization."
        Environmental Ethics 10 (1988): 347-355.
    A wise essay that confronts the apparent conflict in deep
    ecology philosophical positions between the conception of
    nature as the manifestation of an infinite, inexhaustible
    principle of life, beyond the power of humans to harm or
    destroy, and the obligation to preserve and protect nature
    on earth.   If nature in its deepest sense cannot be
    destroyed by human agency, why should we be concerned
    about the local ecosystem?   Freya Matthews sees this
    dilemma as resolved through the doctrine that interconnec-
    tedness is absolutely fundamental to the identity of
    things.   All sentient beings are engaged in self-realiza-
    tion, but the self to be realized is constituted by its
    ecological relationships; consequently, it is in the
    interest of the expanded self that the natural environment
    continue to exist.   Indeed, the author goes beyond this
    to speculate that nature as a whole is a self-realizing
    entity and may be subtly affected by our affirmation.

183 Merchant, Carolyn. "Environmental Ethics and Political
        Conflict: A View from California." *Environmental
        Ethics* 12 (1990): 45-68.
    An analysis of three ethical approaches to the environ-
    ment--egocentric, homocentric and ecocentric--that covers
    their affinities with particular philosophic, social and
    religious ideas and their expression in political con-
    flicts over the protection and use of nature.   According
    to Carolyn Merchant, egocentric ethics focuses on the good
    of the individual and assumes that what is good for the
    individual is good for society.   It is associated with
    capitalism, Hobbesian political philosophy, mechanistic
    science and the Biblical injunction to subdue the earth.
    Homocentric or anthropocentric ethics applies the utili-
    tarian standard of the greatest good for the greatest
    number.   It underlies the resource conservation doctrine
    that developed in the late 19th and early 20th centuries
    and reflects the religious concept of stewardship.   An
    ecocentric ethic is grounded in wholes--ecosystems, the
    planet, the cosmos--and assigns intrinsic value to
    nonhuman nature.   It is influenced by ecology, process
    philosophy and the new physics.   Merchant sees the
    controversy over damming the Stanislaus River as illus-
    trating all three approaches: ranchers and farmers sought
    water for their own use; regulators supported the dam as
    the best use of the river for the general public; and
    protestors tried to protect the natural river for its own
    sake.   This overview article is rich in information and
    insight.   Merchant's distinction between egocentric and
    homocentric, especially, should do much to clarify
    assumptions in controversies over wilderness preservation.

184 Miller, Peter. "Value as Richness; Toward a Value
        Theory for an Expanded Naturalism in Environmental
        Ethics." *Environmental Ethics* 4 (1982): 110-114.
    An effective exposition of richness as characteristic of
    the intrinsic value existing in nature independent of
    human perception.   Richness is composed of resources,
    development and accomplishment, diversity and inclusive-

ness, harmony and integrity, and utility and generativity.
The riches of conscious life are a special case of the
larger context of richness, which includes the "value
potentials and realities of all living things alone and
in interaction with one another and with ourselves" and,
beyond that, of matter and energy. In the closing section
of his paper, Miller explores the implications of his
theory for an environmental ethics that can discriminate
among species on the basis of the extent of their participation in each category of richness.

185   Naess, Arne.   *Ecology, Community and Lifestyle: Outline
       of An Ecosophy.*   Ed. and trans. David Rothenberg.
       Cambridge,   England: Cambridge   University   Press,
       1989.
An English language adaptation of the 1976 Norwegian book
that inspired the deep ecology movement. Naess sets out
in this complex and many-faceted work to relate ecophilosophy (philosophy concentrating on relations to nature)
to the way one lives one's life and to the tactics to be
used in confronting and debating developers and other
opponents of free nature.   In Naess' personal ecophilosophy, termed Ecosophy T, reality is made up of gestalts
whose concrete contents include emotional qualities as
well as qualities usually attributed to objects.   Existence is relational: A is B in relation to C.   The true
self consists of just these multiple relations; and the
key norm in Ecosophy T is Self-realization, the never
completely obtainable goal of learning to identify with
the diversity, complexity, and symbiosis of the world.
Ecosphy T differs from most philosophical positions in
that it is a practice as well as a system.   Self-realization and the joyful experiencing of nature require a
deliberate effort to develop sensitivity to qualities.
The resultant change of consciousness is necessary to
bring about a lifestyle favorable to the "whole ecosphere
in all its diversity and complexity."

186   Naess, Arne.   "The Shallow and the Deep, Long-Range
       Ecology Movement: A Summary."   *Inquiry* 16 (1973):
       95-100.
A summary of an Introductory Lecture at the 3rd World
Future Research Conference held in Bucharest in September
1972. Naess' article is a useful overview in outline form
of the major tenets of the deep ecology movement. It can
serve the reader as a touchstone for interpreting references to deep ecology throughout environmental literature.

187   Norton, Bryan G.   "Conservation and Preservation: A
       Conceptual Rehabilitation."   *Environmental Ethics* 8
       (1986): 195-220.
An argument against the assumption that the terms conservationist and preservationist are synonymous with the
terms anthropocentrist and nonanthropocentrist.   This
mistaken reduction is not only philosophically questionable but practically unwise in that it tends to drive a
wedge between groups whose ecological concerns lead to
similar solutions. The author argues that, in practice,
the significant difference is between those who see only

short-term commercial values in wilderness and those, whether anthropocentrist or not, who value wilderness aesthetically and for its contribution to preserving biological diversity for the long term.

188   Norton, Bryan G.   "Environmental Ethics and Weak Anthropocentrism." *Environmental Ethics* 6 (1984): 131-148.
An earlier formulation of an argument expanded in his 1987 book, *Why Preserve Natural Variety?* Weak anthropocentrism is based on considered preferences, desires or needs that are expressed after careful deliberation. Unlike felt preferences, considered preferences must be consistent with a rationally adopted world view that recognizes fully supported scientific theories of evolution and ecology. The author contends that the considered preferences of a weak anthropocentrism are sufficient to provide protection for wild places and natural environments.

189   Norton, Bryan G.   *Why Preserve Natural Variety?* Princeton, NJ: Princeton University Press, 1987.
An incisive examination of various rationales for the preservation of species in their natural habitats. The author distills the essence of both anthropocentric and nonanthropocentric preservation arguments, including those of Paul Taylor and Holmes Rolston, III, and offers his own expanded version of the anthropocentric point of view based on the recognition of differences in the worthiness of human preferences. Not all felt preferences are equal. Those that can withstand rigorous analysis, for example, the preference for species diversity, should prevail over those that are overly consumptive or shortsighted. Norton also introduces the notion of transformative value. Diverse species and natural ecosystems not only satisfy existing preferences but serve to transform human value systems and lead to questioning materialistic demand values. This is an important book for understanding the existing options for environmental ethics.

190   Partridge, Ernest.   "Nature as a Moral Resource." *Environmental Ethics* 6 (1984): 101-130.
An essay replete with insights from ecology, psychology and sociology brought to bear on the question of whether or not there is a human need to value nature. The author stresses the cognitive and emotional dissonance that results from the basic inconsistency of scientifically recognizing the ecosystemic view of the world, while maintaining the narrow anthropocentric moral view that the natural world has value only as it is useful or entertain-ing for human beings. The moral paradox that we gain happiness and fulfillment only when we cease concentrating on our own gratification and transcend our concern for self applies to our relationships with wild species, ecosystems and landscapes as well as to those with fellow humans. Species narcissism is analogous to individual narcissism. Both lead to emptiness and alienation.

191 Passmore, John.   *Man's Responsibility for Nature:*
        Ecological Problems and Western Traditions. 2nd ed.
        London: Duckworth, 1980.
     A search for sources within the Western traditions on
     which   to   build   a   consensus   for   reducing   pollution,
     conserving resources for future generations, reducing
     population growth, and preserving animal species and areas
     of wilderness.   In an historical excursus John Passmore
     comes to the conclusion that the view of the major Western
     religious and philosophical traditions--Greek and Chris-
     tian but not Hebraic--that everything earthly exists to
     serve man encourages looking at nature as something to
     utilize, not respect.   Other minor traditional strains are
     more supportive.   This is illustrated in the chapter
     "Preservation," where, after demonstrating that there are
     good anthropocentric resource conservation reasons for
     protecting wilderness, he uncovers "seeds" of ethical
     arguments.   These include changing sensibilities in the
     form of concern for animal suffering and reverence for
     life as well as philosophical and religious ideas, such
     as, the Jewish condemnation of wanton destruction and the
     value Aquinas' places on diversity.   In the essay "Atti-
     tudes to Nature" appended to the second edition, he
     proposes conditions for an adequate metaphysics of nature.
     It must not think of "natural processes either as being
     dependent upon man for their existence, as infinitely
     malleable, or as being so constructed as to guarantee the
     continued survival of human beings and their civilisa-
     tion."   The earlier (1974) edition of this book served as
     a   whetstone   to   hone   the   arguments   of   the   developing
     ecocentric ethicists.

192 Reed, Peter.   "Man Apart: An Alternative to the Self-
        Realization Approach."   *Environmental Ethics* 11
        (1989): 53-69.
     A philosophical explanation for the sense of awe that many
     persons feel in the presence of wild nature.   As an
     alternative to theories of the relationship of man to the
     natural world that emphasize human beings as a part of the
     community of life--deep ecology, for example--the author
     proposes a conception of nature inspired by Martin Buber's
     I-Thou   model   and   Rudolph   Otto's   idea   of   the   Holy.
     Nature's intrinsic value consists of its otherness and
     independence of self.   The value of wilderness is directly
     intuited and experienced as wonder in the face of a
     landscape that is empty of human artifacts.   For Reed the
     "myth of progress is the myth of Narcissus: it merely
     brings us back to ourselves" and the tedious boredom of
     a   man-made   world.   He   recognizes   the   philosophical
     weaknesses of an intuitionist ethic; yet he   convincing
     in his contention that an adequate environmental ethic
     must take account of the role of emotion in wilderness
     appreciation.

193 Regan,   Tom.   "The   Nature   and   Possibility   of   an
        Environmental   Ethic."   *Environmental   Ethics*   3
        (1981): 19-34.
     A tentative but logically crafted preparation of the
     ground for creating an environmental ethic.   The author,

known for his animal rights position, accomplishes three things in this article: he advances what he is convinced are two necessary conditions for an environmental ethic; he examines and finds philosophically unsound the arguments against the need for a separate environmental ethic to assure the preservation of wilderness and wildlife; and he elucidates the meaning of inherent goodness. The conditions Regan considers necessary are that nonhuman beings have moral standing and that the class of beings with moral standing be larger than the class of conscious beings.

194 Rodman, John. "The Liberation of Nature." *Inquiry* 20 (1977): 83-131.
A consideration of the need for untamed others not under the control of humankind as the foundation for valuing wilderness and wild things. John Rodman uses the occasion of reviewing books by Peter Singer and Christopher Stone to present his own view that theories that extend legal or moral rights to nonhuman beings perpetuate the homocentric bias of resource conservation arguments, albeit in a subtler, more compassionate way. Rodman terms this the "Method of Argument from Human Analogy and Anomaly." Though well-intended, such progressive extension ethics legitimize the "basic project of modernity--the total conquest of nature by man." Everything is drawn into the human orbit to be managed and controlled. What is required instead is an ethic based on a larger order of which humans are a part. Only such an ethic can liberate diversity, both human and natural, from "totalitarian monoculture." Rodman's essay serves as an exciting introduction to some seminal ideas about the relationship between the oppression of nature and the oppression of human potential.

195 Rolston, Holmes, III. *Environmental Ethics: Duties to and Values in the Natural World*. Philadelphia, PA: Temple University Press, 1988.
A mature formulation of the author's ethical insights and arguments developed over two decades of writing and debate. In this volume Rolston presents his model of an environmental ethic sequentially, examining first duties to humans and higher animals, then duties to all organisms, next duties to endangered species and, finally, duties to whole ecosystems, as the generators of all individuals and species. Humans have obligations at each of these levels of organization, but the duties to ecosystems are critical in that without healthy ecosystems organisms die off and the process of evolutionary speciation stops. Unlike some environmental ethicists, Rolston leaves room for a hierarchy of intrinsic value among species, which accords humans superiority because they are capable of a greater range of values than any other known organism. Paradoxically, this superiority is only manifest when humans achieve an altruistic world view that leads to the protection of other species and the preservation of ecosystems.

196 Rolston, Holmes, III. *Philosophy Gone Wild*. Buffalo, NY: Prometheus Books, 1986.
A collection of the author's articles and essays reprinted from journals and anthologies and organized in four sections: Ethics and Nature; Values in Nature; Environmental Philosophy in Practice; and Nature in Experience. There is an advantage in having these pieces brought together in that the reader is able to follow Rolston's treatment of his major themes in varied styles and contexts over a period of nearly twenty years. His probings of the cultural underpinnings of the doctrine of the naturalistic fallacy and his conviction that wild nature is both a generator and a repository of values appear as technical philosophic argument, as practical guidance for business people and government officials, and as personal response to solitude by a wilderness lake. The result is a fuller understanding of the intrinsic value of wilderness than could have been conveyed by sophisticated argument alone.

197 Rolston, Holmes, III. "Values Gone Wild." *Inquiry* 26 (1983): 181-207.
A defense of intrinsic value in the natural world presented in the form of a journey into the wilderness. Holmes Rolston, III acts as a wilderness guide pointing out along the way the values generated in wild places. In his company the reader becomes aware of the "wild generative roots" that form human beings, the wild creatures akin to us and those whose differences make them appear alien. Wilderness is the womb of value bringing into being over time the "storied achievement" of an ever-changing complex of things good in themselves and good in their niche. Wilderness areas--the small percentage of earth left unpaved and undeveloped--are precious precisely because they are not resources with instrumental value for humans. Travelling in them dramatizes the artificial narrowness of our conventional values and brings us as near as we can come in phenomenal experience to the primal ground and the prime mover. As a philosophical work that is yet rich in natural imagery and personal feeling, Rolston's essay makes a special contribution to environmental ethics. It is reprinted in the author's anthology *Philosophy Gone Wild*.

198 Rolston, Holmes, III. "Valuing Wildlands." *Environmental Ethics* 7 (1985): 23-48.
A telling argument against the use of cost-benefit analysis in making public land use decisions about wild nature. The author's taxonomy of wilderness values summarizes in a few pages the various attitudes towards wilderness that have been expressed in the literature. He, then, uses these values as the basis for developing a series of decision-making rules to substitute for evaluations using cost or price standards. This is an excellent place to begin an overview of wilderness philosophy. It is reprinted in the author's book *Philosophy Gone Wild: Essays in Environmental Ethics*.

199 Routley, Richard, and Val Routley. "Human Chauvinism and
       Environmental Ethics." *Environmental Philosophy.*
       Ed. Mannison, D. S., M. A. McRobbie and R. Routley.
       Canberra, Australia: Australian National University,
       1980: 96-189.
     A closely argued rejection of the dominant Western ethical
     system, termed human chauvinism, or any of its milder
     variations--stewardship, cooperation in perfecting nature,
     or resource conservation--as possible foundations for an
     adequate environmental ethic. All Western systems are
     based on a conceptually indefensible sharp moral separa-
     tion between man and all other species. Satisfactory
     divisions of moral relevance can be made, but none of them
     results in the isolation of human beings as a unique
     privileged class. Moreover, the location of all value in
     humans is counter-intuitive. Wilderness advocates often
     defend the preservation of wild areas on anthropocentric
     grounds as the source of an aesthetic, quasi-religious
     experiences in human beings; but most would be appalled
     at substituting a similar experience produced by a
     hypothetical wilderness machine. They correctly intuit
     that the value they experience resides, in some sense, in
     the wilderness itself. This challenging essay is fre-
     quently referred to and cited in the literature.

200 Shrader-Frechette, Kristin. "Four Land Use Ethics: An
       Overview." *The Environmental Professional* 9 (1987):
       121-132.
     An introduction to emerging ways of relating to the land
     based on the author's keynote address at the 1986 meeting
     of the National Association for Environmental Education.
     Kristin Shrader-Frechette outlines the assets and liabili-
     ties of each of four land use ethics proposed as solutions
     to the contemporary crises of unequal distribution,
     spillover from exploitative land use practices, degrada-
     tion of the land as a biotic community, and the inadequacy
     of legal protection for the land. Land reform appears
     irrelevant to issues of wildlife and wilderness preserva-
     tion; but the three remaining ethics--land use-ethics,
     land-community ethics, and land-rights ethics--could serve
     to protect natural ecosystems. Land-use ethics would
     place controls on the rights of property owners to use
     their land in any way they desired regardless of the
     effects on broad human interests. Land-community ethics
     and land-rights ethics are ecocentric rather than anthro-
     pocentric and introduce expanded concepts of moral
     obligation. This is a useful overview of environmental
     ethics. However, the treatment of the assets of the
     ecocentric positions should be supplemented with reading
     from the appended bibliography.

201 Stone, Christopher. *Earth and Other Ethics: the Case
       for Moral Pluralism.* New York: Harper & Row, 1987.
     A complex and challenging argument by a law professor for
     the development and adoption of an ethical pluralism as
     the proper medium for making ethical judgments about
     complex moral situations, especially, those involving
     individual animals, species, ecosystems and inanimate
     entities, as well as persons. The predominant moral

philosophies are monisms which fail to account for
different kinds of ethical activities, for example,
evaluating personal choices, evaluating choices for
others, and judging character; moreover, they are unable
to deal with the diversity of entities that have a claim
to moral consideration. Pluralism conceives the world of
morals as partitioned into separate though related planes
where different kinds of moral arguments and intuitions
prevail. On some of these planes the reasoning will be
more closely related to aesthetic judgments than to the
traditional either-or formulations of ethics. Such moral
pluralism enables us to consider the well-being of
nonpersons without according them the status of rights-
holders. Stone's book should prove a stimulus to new
approaches in the development of environmental ethics.

202   Stone, Christopher D. "Moral Pluralism and the Course
    of Environmental Ethics." *Environmental Ethics* 10
    (1988): 139-154.
A restatement of the argument, introduced in the author's
Earth and Other Ethics, that abandoning the metaphysical
and metaethical assumption that all types of moral
judgments must be based on a single ethical system would
be a fruitful direction for environmental ethics. The
development of different ethical systems for the land, for
higher animals, for kinship relations, for political
action, etc., could bring unexpected benefits, including
the possibility that an action might be discovered to be
justified or right according to several systems or that
certain actions are morally unacceptable in all systems.

203   Taylor, Paul. *Respect for Nature: A Theory of
    Environmental Ethics.* Princeton, NJ: Princeton
    University Press, 1986.
A step by step, logically argued attempt to establish
rational grounds for a system of moral principles to
govern human treatment of animals and their wild communities. The result is a biocentric environmental ethic that
expresses the ultimate attitude Paul Taylor calls respect
for nature. Respect for nature entails regarding all
living wild things as having inherent worth and deeming
"their wild existence as worthy of being preserved as an
end in itself and for their sake." It also involves
taking practical steps to preserve natural ecosystems and
feeling pleasure in their continued existence. In
Taylor's biocentric ethic human beings, as moral agents,
are obliged to consider other beings as moral subjects
with their own legitimate goals, independent of their
usefulness in furthering human ends. This environmental
ethic neither supplants nor supersedes the legal and
ethical systems that govern relations among humans.
Conflicts may, in fact, occur between these two ethical
systems. In an interesting concluding chapter, the author
discusses workable principles for dealing with competing
claims of duties to humans and duties to nonhumans. This
book is one of a handful of essential texts for studying
wilderness philosophy and for developing a personal
wilderness ethic.

204   Thompson, Janna L. "Preservation of Wilderness and the
      Good Life." *Environmental Philosophy: A Book of
      Readings.* Ed. Robert Elliot and Arran Gare.
      University Park, PA: Pennsylvania State University
      Press, 1983.
      An interesting essay that asks the question of whether,
      if humans can sustain themselves in comfort in a complete-
      ly tamed world, there are any grounds for preserving
      natural ecosystems and wilderness. Answering this
      question in the affirmative assumes a revolution in
      ethics. Such a revolution is plausible if we go beyond
      the arguments of academic ethical philosophy to take
      account of "world views, empirical hypotheses and social
      analyses." The change in ethical consciousness that is
      actually occurring can be understood in the context of
      contemporary biological and ecological science and recent
      ideas about social existence. Janna Thompson argues that
      the views of Marcuse and some other radical thinkers
      provide a rationale for preserving wilderness. Domination
      of one group of people over another is now recognized as
      unjust in much of the world. The master-slave relation-
      ship of man and nature is linked to the old ethic of
      domination in the social world. If the exploitative and
      centrally controlled technologies characteristic of
      dominance are abandoned, humans can realize their creative
      potential in ways that are consistent with respect and
      appreciation for natural systems.

205   Van De Veer, Donald, and Christine Pierce. *People,
      Penguins, and Plastic Trees: Basic Issues in
      Environmental Ethics.* Belmont, CA: Wadsworth, 1986.
      An anthology designed as a text for college students in
      environmental ethics courses. Articles were chosen on the
      basis of their relevance to the questioning of traditional
      anthropocentric values and the search for more comprehen-
      sive value systems now taking place in philosophy. The
      collection is arranged in thematic sections, each preceded
      by the editors' discussions of issues and controversies.
      The section entitled "Wilderness: What Is It Good For"
      contains pieces by Rene Dubos and Robert Eliot; but some
      of the articles in other sections are equally relevant,
      for example, Mary Midgley's "Duties Concerning Islands"
      or Paul Taylor's examination of respect for nature.

206   Vest, Jay Hansford C. "The Philosophical Significance
      of Wilderness Solitude." *Environmental Ethics* 9
      (1987): 303-330.
      An interesting definition of wilderness solitude arrived
      at through etymological analysis. The author criticizes
      the Forest Service's dictionary interpretation of solitude
      as aloneness, which has lead to refusals to designate
      prairie-like wildlands as wilderness areas because they
      lack vegetative screening. The meaning of solitude, when
      applied to wilderness is soul-mood, the numinous experi-
      ence of at-one-ment with the natural world.

207   Warren, Karen J. "The Power and the Promise of
      Ecological Feminism." *Environmental Ethics* 12
      (1990): 125-146.

An argument for recognizing the connection between the
domination of women and the domination of nature in
Western culture in structuring an adequate environmental
ethic for our time.   In the patriarchal conceptual
framework women and nature are labeled physical and placed
beneath men and the rational in the hierarchy of values.
A valid environmental ethic must also be a feminist ethic
because both women and nature are degraded and exploited
by the culturally powerful "logic of domination" that
justifies oppression on the basis of hierarchical thinking.   Ecofeminism denies the logic of domination and
chooses instead to ground ethical discourse and practice
in the patterns that emerge from diverse voices in varying
historical and cultural contexts.   How this might work in
particular cases is illustrated by the method of first-
person narrative, which Karen Warren recommends as a
vehicle for refocusing decision-making and theory building
in environmental ethics.   As her account of a rock
climbing experience demonstrates, personal history can
document values that are often overlooked in traditional
argument and can provide unique insights into our relationships with nonhuman nature.

POETRY
AND FICTION

208 Abbey, Edward. *Hayduke Lives!* Boston, MA: Little,
      Brown, 1990.
A sequel to *The Monkey Wrench Gang*, in which Edward Abbey
brings back the gang members and the mysterious Lone
Ranger, older but not subdued, to aid youthful Earth
First!ers in destroying GOLIATH, the G.E.M. or giant earth
mover of Arizona, and prevent their old adversary Bishop
Dudley Love and his corporate backers from Syn-Fuels from
turning the canyon country into a job-generating nuclear
mine. Abbey's satiric wit and bawdy humor give his
narrative a comic cast. There is much laughter in this
book; but the underlying conflict between lovers of
"spareness, beauty, open space, clear skies" and the
dominant culture of growth and progress is dead serious.
Bishop Dudley exhausting his aging body in an attempt to
conceive his twelfth child or waxing sentimental over the
motel and spa he envisions on a bulldozed mesa is a figure
of fun; yet the megamachine of techno-industrial civilization applauds his values. For the eco-warrior there is
only "public obloquy." "Congressmen will fulminate,
senators abominate, bureaucrats denunciate and all the
vipers of the media vituperate." The Monkey Wrench Gang
succeeds in their mission, but the "planetary empire of
growth and greed" remains firmly in control. This
guerrilla novel is a worthy culmination to Abbey's work.

209 Abbey, Edward. *The Monkey Wrench Gang*. Salt Lake
      City, UT: Dream Garden Press, 1990.
An illustrated reprint edition of Edward Abbey's rollick-
ing, iconoclastic novel of eco-sabotage published in 1975.
Abbey's outraged gang members take on the juggernaut of
expanding military-industrial civilization, blowing up
bridges and dismantling bulldozers, in an attempt to save
the desert wilderness of Utah and the Southwest. Abbey
covers all the exploitative villains of western expan-
sion--the clearcutting lumber companies and their Forest
Service allies, the power companies strip mining coal from
Black Mesa, the development boosters busy turning the
wilderness into a drive-through tourist attraction. These

are parts of what gang member Doc Sarvis calls the "mad machine." But, for all their insouciance and foolhardy courage, the gang can only annoy not destroy the powerful enemy. The novel ends with an uneasy truce and a hint of more adventures to come.

210 Daniel, John. *Common Ground: Poems by John Daniel*. Lewiston, ID: Confluence Press, 1988.
Poems that radiate wonder and quiet joy at the existence of the natural world. John Daniel celebrates the perfection of wild things just as they are in themselves--a curious raccoon peering through a lamplit window, "curiously sure;" an owl sensing the night "with a clarity we could not bear;" joshua trees rooted in their own "promised land." He is intrigued by the mystery that the earth not only exists "but is beautiful." In the title poem "Common Ground" he imagines all living things but especially humans, who are gifted with clear consciousness, as the planet awakening to sentience, seeing, hearing touching and tasting. His own poems, filled with imagery of sight and sound and motion, might be understood as exemplars of this marvelous process.

211 Eiseley, Loren. *Another Kind of Autumn.* New York: Scribner's, 1977.

212 ---. *The Innocent Assassins.* New York: Scribner's, 1973.

213 ---. *Notes of an Alchemist.* New York: Scribner's, 1972.
Three collections of poems by a well-known scientist. Eiseley writes in spare, unrhymed verse of the eternal themes of poetry--the past, the passing of time and the impermanence of things, except his past contains all of evolutionary time and his impermanence the passing of eons, the end of civilizations, and the extinction of humankind. He has a unique ability to imagine and communicate the consciousness of animals and the closeness of humans to the animal world. In "The Leaf Pile" the "apple smell of autumn" awakens the old limbic brain, and he sees his ancient snout, scaled and ugly, hiding among the leaves. "In the Fern Forest of All Time I Live" finds him staring at his human hand resting on his scholar's desk and seeing it "present with the gorgosaurs," the appendage of a timeless reptile waiting out the demise of man. Other poems express his sorrow, as he looks back in old age, for the passing of the wild country of the pioneers and of his own youth. The smooth lawns and the smooth faces of an endless suburbia, where there is no place for a "boy or a badger to hide in the hedgerows," bring a feeling of unredeemable loss. Like all lovers of the wild, Eiseley questions what we have gained from our destruction of beauty and diversity.

214 Faulkner, William. "The Bear." *Go Down Moses.* 1942. New York: Vintage, 1973.
A literary masterpiece exploring the significance of wilderness in American culture. William Faulkner evokes, through the image of an almost mythical great bear and the story of a boy's growth to manhood under the tutelage of

an old hunter, son of a Chickasaw chief and a Negro slave, both the numinous quality of the wilderness experience and a wrenching sense of loss at the passing of the wild land. For Faulkner's hero, Isaac McCaslin, this pristine continent should have been held "mutual and intact in communal anonymity of brotherhood;" instead, we allowed it to become tainted and accurst, turning it into squares and rectangles to be exploited and sold, and the people on it into slaves to be bartered and traded for cash. But before the last unaxed woods is leased to the loggers, young Ike is given the opportunity to learn the meaning of wilderness. Leaving his gun behind and, at the last, even his compass and watch, accepting fear yet not being afraid, he watches the wilderness coalesce in Old Ben, the big bear "dimensionless in dappled obscurity."

215   Haines, John.   *News from the Glacier: Selected Poems 1969-1980*.   Middletown, CT: Wesleyan University Press, 1982.
Poems grounded in evolutionary time and rooted in the Alaskan landscape. John Haines writes of Alaskan weather and wildlife, mixing accurate observation with a sense of the mystery of other lives.   His wolves and owls and caribou are both flesh and blood animals struggling to survive in a cold climate and mythic icons, looming gray or white against the polished ice of the arctic moonlight. Throughout the collection he conveys the strength of this spare, glaciated land to resist taming by a restless generation stuck in its own "desolate time."   As he looks back thirty years in the poem "Homestead" to his first days building a wilderness life at Richardson Hill, he recalls tales he heard of earlier settlers and seekers of fortune, who left no more record of their passing than the plopping of a few of berries "tumbled in a miner's pail."

216   Jeffers, Robinson.   *The Beginning and the End, and Other Poems*.   New York: Random House, 1963.
Later poems of Robinson Jeffers, collected after his death in 1962.   In these mostly short poems Jeffers continues to express, sometimes in irony, anger or resignation but always with passionate honesty, his lifelong conviction that value resides in nature as a whole.   His loved subject remains what it has always been--the organic universe of "Mountain and ocean, rock, water and beasts and trees." These are the actors; the function of humanity is to interpret them in symbols.   Human beings are sapped of their dignity when, through explosive population growth and violent technology, they destroy the integrity of the planet, substituting their "stupid dreams and red rooster importance" for the reality of the natural world.   In contrast to the confused dreams of the ego, Jeffers evokes the high, clear beauty of God made visible in "the cliffs, the ocean, the sunset cloud."

217   Jeffers, Robinson.   *The Selected Poetry of Robinson Jeffers*.   New York: Random House, 1938.
Some of the earliest and most poetically intense modern expressions of evolutionary and ecological wholeness in literature.   Robinson Jeffers uses imagery from the spare

land and seascapes of the then still pristine Monterey coastal mountains to write of elemental human emotions and permanent things--sea, rock and sky and the splendor of the organic universe. Often he contrasts these with the detritus and ephemera of "Progress" and the triviality of mass populations insulated in crowded cities. The poems in this collection were chosen by Jeffers himself and are arranged chronologically according to the order in which they were published. Those most relevant to wilderness values tend to appear in the second half of the book. Among the many notable poems, "The Answer" deserves special mention as the source of the famous lines that admonish us not to love man apart from the "wholeness of life and things."

218   Snyder, Gary. *Turtle Island.*   New York: New Directions, 1974.

A collection of poems and occasional prose pieces organized around the idea of the American continent--Turtle Island--as a community of ecosystems rather than a set of arbitrary political units. In the essay "Wilderness" the author explores the idea that a government truly representative of the reality of the planet would incorporate the interests of the "creeping people, and the standing people, and the flying people, and the swimming people" into its councils. The poems themselves repeatedly contrast the living, the natural, and the wild with the mechanical, artificial, and the economic. Gary Snyder is particularly effective in conveying the grossness and violence of our relationships to the nonhuman world: a bulldozer grinds and slobbers, sideslips and belches and tosses aside the "skinned-up bodies of still-live bushes." This volume also reprints "Four Changes," a prescription for a sustainable society first published in 1969.

RESEARCH STUDIES

219 *Americans Outdoors: The Legacy, the Challenge.* Report
of the President's Commission on Americans Outdoors.
Washington, DC: Island Press, 1987.
A successor to the Outdoor Recreation Resources Review
Commission (ORRRC) report of 1962, which "helped launch
the modern environmental movement." This report is a
consensus document, quoting input from diverse experts,
interest groups, organizations, and citizens and referenc-
ing the results from a nationwide telephone survey of
2,000 people sponsored by the National Geographic Society.
As such, its proposals are less ambitious than those
espoused by the conservation community. In general, the
Commission found that Americans are still in love with the
land and wish to preserve the outdoors for their children's children. They also found that we are "facing a
deterioration of the natural resources base, and of the
recreation infrastructure." As nearby woods, wetlands,
rivers and beaches are dammed, paved and developed,
opportunities to experience nature are reduced for our
increasingly urban population. Though the report contains
several sections on wilderness topics, the emphasis is on
preserving greenways in built areas. The report also
recognizes the need for conservation education, a strong
conservation ethic and some method of taking recreation
and ecological values into account when land use decisions
are made if government, private industry and citizens
groups the Commission recommends is to succeed.

220 *The Continental Group Report: Toward Responsible
Growth; Economic and Environmental Concern in the
Balance.* Stamford, CT: The Continental Group, 1982.
A comparative opinion study of the economic and environmental attitudes of 1,300 members of the general public,
263 corporate executives and 343 members of environmental
groups across the country, as well as five case study
groups. Overall, the study reveals that a large percentage of Americans feels that both economic growth (80%) and
protecting nature (77%) should be given high priority.
Nevertheless, some interesting differences exist among

groups regarding the relative importance of utilizing versus preserving natural resources. In general, the public falls between the executives and the environmentalists in their willingness to protect the environment at the cost of economic growth. Younger, more highly educated and less religiously committed persons put higher value on preservation than do their opposites. The report is a valuable source of evidence for changing attitudes toward wilderness and wildlife protection.

221 *Wilderness and Recreation--A Report on Resources, Values, and Problems.* Outdoor Recreation Resources Review Commission Report No. 3. Washington, DC: Outdoor Recreation Resources Review Commission, 1962.
A report to the ORRRC by the University of California Wildlife Research Center, which was considered by the ORRRC in its own report to the President and Congress in January, 1962. For the student of wilderness values, the most relevant section of the report is the survey of the appeals of wilderness to users in the Boundary Waters Canoe Area, Mount Marcy in New York, and the High Sierras. In all three areas the aesthetic-religious appeal of observing the beauty of nature was found to be the most frequently expressed value. Nearly equaling this were the "exit-civilization" values. Most wilderness users appreciated the contrast wilderness provides to stressful, crowded urban life. The authors of the report surmise that wilderness may make an "important contribution to the mental health of those who use it."

222 Dolin, Eric Jay. "Black American Attitudes toward Wildlife." *Journal of Environmental Education* 20.1 (1988): 17-21.
A review of the limited body of research concerning the attitudes of Black Americans toward wildlife and the natural environment. In addition to two studies, one in 1976 of 70 residents in the black area of Denver and the other part of the U.S. Fish and Wildlife Service survey of over 3,000 Americans between 1978 and 1981, Dolin reports the results of a few informal surveys on college campuses. None of these show that the natural environment is a major concern for African Americans. The contrast between blacks and whites is most pronounced at higher levels of education and income. After briefly discussing the various theories that have attempted to account for this difference, Dolin concludes that without additional study and more carefully designed surveys meaningful explanation is impossible.

223 Driver, D. L., Roderick Nash, and Glen Haas. "Wilderness Benefits: A State-of-the-Knowledge Review." *Proceedings--National Wilderness Conference: Issues, State-of-Knowledge, Future Directions.* Fort Collins, Colorado. 23rd-26th July 1985. Ed. Robert C. Lucas. General Technical Report A 13.88: INT-220. Ogden, UT: Intermountain Research Station, 1987: 294-319.

A taxonomy of wilderness benefits based on interpretations of wilderness user research and nature writing. The authors define a benefit as a "specific improved condition or state" of an individual or society or nonhuman organism. Benefits are clustered in three major categories according to their recipients: personal benefits, social benefits and inherent/intrinsic benefits, the least anthropocentric type. Within these categories benefits are further classified to give a conceptual overview of wilderness values. The authors' purpose in constructing this array is to isolate those benefits that are central to wilderness and, so, can constitute the basis for a wilderness philosophy and provide guidelines for wilderness management. They identify six core values: preservation of representative natural ecosystems and maintenance of species diversity; spiritual values that "capture themes of natural cathedrals; aesthetic values that go beyond pretty scenery; inherent or intrinsic values that "hypothesize that non-human organisms have a right to exist;" and specific types of recreation that require a wilderness setting.

224 Haas, Glen E., Eric Hermann, and Richard Walsh. "Wilderness Values." *Natural Areas Journal* 6.2 (1986): 37-43.
A survey of the Colorado public, both wilderness users and non-users, asking people to rate the importance of thirteen different wilderness values. Although the results are tentative because of the low rate of return, the responding populace was representative of the residents of the state in terms of income, education, age, size of family, and geographic distribution. The findings have importance for wilderness management in that they show recreation to be significantly less of a concern for the public than protecting wildlife habitat or water or air quality.

225 Hammitt, William E. "Cognitive Dimensions of Wilderness Solitude." *Environment and Behavior* 14 (1982): 478-493.
A presentation of a conceptual framework based on information-processing theory for investigating the meaning of wilderness solitude together with an analysis of an initial study conducted to test its usefulness. The author hypothesizes that the solitude valued by wilderness users is not complete isolation from contact with others but a more complex phenomenon characterized by freedom of choice and absence of stress in information processing. Wilderness users seek the privacy of being in an intimate, small group social setting where they can control the extent of interaction and communication. They also value the pristine natural environment, in which inherently fascinating sights and sounds can be processed in the state of relaxed alertness termed "involuntary attention." A 1981 test survey of 109 wilderness campers--college students enrolled in outdoor recreation or natural resources classes and members of hiking clubs--tended to confirm the explanatory value of Hammit's model.

226 Kaplan, Rachel, and Stephen Kaplan. *The Experience of Nature: A Psychological Perspective.* New York: Cambridge University Press, 1989.

A review and analysis of the research conducted since 1979 by the authors and other psychologists on the relationship between people and the natural environment. Part I demonstrates unequivocally the strong and consistent preference humans show for natural over man-dominated environments. When asked to rank photographs of actual scenes, subjects overwhelmingly rate built environments significantly lower than scenes combining trees and open space. Rachel and Stephen Kaplan interpret this preference as an expression of an underlying human need for surroundings suited to the species. In Part II they report on ten years of research with the Outdoor Challenge Program to assess the satisfactions and benefits people derive from wilderness and the equally positive effects of nearby nature. Initially, they expected wilderness benefits to emphasize acquisition of outdoor skills and competencies; they discovered, however, that what participants in the program valued most was the sense of peace and wholeness they experienced as they learned to participate in their wild surroundings. This role of nature as a restorative environment is discussed theoretically in the concluding section. The Kaplans' extensive research suggests a scientific foundation for the intuitive conviction of nature writers that the wilderness experience contributes to personal and spiritual growth.

227 Kaplan, Stephen. "Aesthetics, Affects, and Cognition: Environmental Preference from an Evolutionary Perspective." *Environment and Behavior* 19.1 (1987): 3-32.

A review and analysis of the ongoing psychological research concerning the environmental preferences of human beings. Since 1972, when he first identified complexity, mystery and coherence as variables in landscape preferences, Stephen Kaplan has been investigating the possibility of an evolutionary basis for the overwhelming human preference for natural landscapes over built environments. Study after study gives results that are compatible with the evolutionary hypothesis. The preference for natural scenes is immediate and certain, indicating the presence of unconscious cognitive processes linked with positive affect. Among natural scenes, subjects prefer those that suggest support for human life. Typically such scenes combine mystery and coherence and thus promote both understanding and new learning. Water, trees and foliage, elements that characterize environments where human survival is likely, evoke strongly positive responses. Kaplan concludes that aesthetic preferences "reflect neither a casual nor trivial aspect of human makeup." There is an obvious significance for wilderness preservation in these studies.

228 Kellert, Stephen. *Public Attitudes toward Critical Wildlife and Natural Habitat Issues.* I 49.2: At 8/phase 1. Washington, DC: U. S. Fish and Wildlife Service, 1979, c1982.

The analysis of a survey conducted in 1978 to determine the attitudes of the American people towards wildlife issues. The study was designed and conducted for the FWS by the author with the assistance of Joyce Berry under the auspices of the Yale School of Forestry and Environmental Studies. The survey instrument required that respondents weigh wildlife values, such as, protecting endangered species or preserving wilderness habitats, against resource development projects of varying social or economic benefit. Except for a special random sample of livestock producers and trappers, who were mailed ques- tionnaires, the subjects' opinions were gathered in lengthy interviews. The results demonstrated that the general public is far more willing to curtail economic growth in order to protect wildlife and wilderness than is usually assumed. The single most significant variable in separating pro-wildlife attitudes from exploitative attitudes was education. The more highly educated the more willing people are to accept limits on human activity to preserve wild species and habitat. Somewhat surpris- ingly, Alaskans, regardless of education, indicated this same willingness. The only persons consistently opposed to promoting wildlife and wilderness causes were sheepmen and ranchers.

229   Kellert, Stephen R. "Assessing Wildlife and Environmental Values in Cost-Benefit Analysis." *Journal of Environmental Management* 18 (1984): 355-364.
A proposal for a system of benefits measurement that would satisfy the "tyranny of numbers" that tends to govern our society, while still ensuring that intangible and qualita- tive wildlife and wilderness values are given appropriate consideration in political decision-making. Even when law directs that non-consumptive values be taken into consid- eration, the fact that the only numbers used are dollars creates a bias in favor of commodity values in environmen- tal controversies like that over Tellico dam. Stephen Kellert argues that it is possible to develop a universal non-monetary value unit that can "facilitate some kind of additive numeric evaluation." As a first step towards this, it would be necessary to specify all our wilderness values--for example, naturalistic or outdoor, ecological, aesthetic--so that in any particular case it becomes possible to categorize what we are talking about. The second and most difficult task is to develop a system for assigning numbers to these value units. Kellert suggests that the scales used in his earlier study of American attitudes towards wildlife might provide a basis for further research.

230   Lucas, Robert C., comp. *Proceedings--National Wilderness Research Conference: Current Research.* Fort Collins, Colorado. 23rd-26th July 1985. General Technical Report A 13.88:INT-212. Ogden, UT: Intermountain Research Station, 1986.
A collection of research papers presented at a conference on wilderness management sponsored by interested agencies, including the Forest Service and the National Park Service. The papers are organized into ten sections, some

of them, such as, wilderness fire research and wilderness air quality research, quite technical. The section entitled "Wilderness Benefits Research" contains several pieces relevant to wilderness values. In addition to those focused on developing measures for non-economic values, some few touch on psychology and religion. Robert A. Young and Rick Crandell attempt to assess the self-actualization value of extended wilderness use, and Susan P. Bratton contributes an interesting linguistic analysis of the meaning of wilderness in the four Gospels. She concludes, contrary to received opinion, that the Gospels portray wild nature as a "place of spiritual encounter, both divine and satanic." The wilderness fish and wildlife research section contains many affirmations of a biocentric position that attributes intrinsic value to wild species. Overall, the collection serves to illustrate the broad range of interests and disciplines engaged in wilderness research.

231   Lucas, Robert C., ed. *Proceedings--National Wilderness Conference: Issues, State-of-Knowledge, Future Directions.* Fort Collins, Colorado. 23rd-26th July 1985. General Technical Report A 13.88: INT-220. Ogden, UT: Intermountain Research Station, 1987.
One of two volumes of the proceedings of this conference. The collection begins with a section devoted to perspectives on wilderness values and wilderness management followed by state-of-knowledge reviews of research on wilderness resources and wilderness users and concludes with a series of brief speculations on future directions. The detailed reviews of research are indispensable for study of wilderness topics. Those covering attitudes toward wilderness and wilderness benefits are annotated separately in this bibliography. Several essays in the first section are also highly relevant to wilderness values, in particular, Harry Crandell's "Congressional Perspectives on the Origin of the Wilderness Act and Its Meaning Today" and George H. Stankey's "Scientific Issues in the Definition of Wilderness." Throughout, the collection offers hints of a shift in wilderness values away from recreational use and towards long-term ecosystem preservation. Wilderness managers appear increasingly concerned that accommodating pressures for recreational use will undermine the purposes of the Wilderness Act and result in areas that are wilderness in name only.

232   Milbraith, Lester W. *Environmentalists: Vanguard for a New Society.* Albany, NY: State University of New York Press, 1984.
An exhaustive analysis of the results of a survey of changing environmental belief systems in the United States, Germany, and England. The survey was undertaken in 1980 and repeated for reliability in 1982. The author, heading a U. S. team supported by the Environmental Studies Center of the State University of New York at Buffalo, joined scholars from European countries to design the project and to devise questions that would assess the degree of commitment on the part of various groups of respondents to the dominant social paradigm (DSP) versus

the new environmental paradigm (NEP). The results show a basic distinction between two contending groups of nearly equal size at either end of the spectrum of opinion--the rearguard and the vanguard--in what they believe is the proper relationship of man to nature. The vanguard, younger, more female, highly educated, and containing a large proportion of environmentalists, values nature and believes it should be preserved for its own sake; the rearguard, containing a high proportion of business, government, and media elites, views nature as a resource for producing goods. The majority of respondents fall somewhere in-between but sympathize more closely with the vanguard on many environmental issues. Overall the surveys reveal that a change in values that rates environmental protection above economic growth is taking place. Detailed appendices include the original survey forms and explanations of statistical methods and fieldwork procedures.

233   Rossman, Betty B, and Z. Joseph Ulehla. "Psychological Reward Values Associated with Wilderness Use: A Functional Reinforcement Approach." *Environment and Behavior* 9.1 (March 1977): 41-66.
A study based on the social learning theory which posits that persons select behaviors and environments according to the variety and importance of expected rewards and the degree to which these rewards cannot be obtained in other environments. To assess the social value of wilderness the authors asked people what rewards they expected to gain by visiting the wilderness, how important these rewards were to them, and whether such rewards could be obtained in a built environment as well. A total of 94 college students at the University of Denver completed a questionnaire which asked them to rate the importance of 30 benefits available in the wilderness and the likelihood of obtaining these in various other environments. In general, the results showed that there are valued rewards related to tranquility, natural beauty and escape from stressful urban life that are both important and felt to be uniquely obtainable in a natural environment. The authors suggest that the loss of natural areas to urban and industrial development could have serious social-psychological costs, including a possible increase in the use of mind-altering drugs or maladaptive personal withdrawal from one's surroundings. Further research is recommended.

234   Scott, Neil. "Toward a Psychology of Wilderness Experience." *Natural Resources Journal* 14.2 (1974): 231-237.
A psychiatrist's ruminations on the question of whether there is a "positive mental health value in non-degraded environments." The article briefly summarizes passages from the writings of famous naturalists--Catlin, Muir, Eiseley, Leopold and others--that recount powerful psychological experiences of altered states of consciousness while in a wilderness setting. Neil Scott suggests that such experiences meet the criteria for peak experiences of self-actualization as described by psychologist

Abraham Maslow. His expectation is that properly designed research would show that "wilderness experiences are more likely to foster self-actualization and the occurrence of peak experiences than outdoor activity in more degraded environments." If that is so, he argues, loss of wilderness would entail a loss of potential for human development.

235 Stankey, George H., and Richard Schreyer. "Attitudes Toward Wilderness and Factors Affecting Visitor Behavior." *Proceedings--National Wildlife Research Conference: Issues, State-of-Knowledge, Future Directions.* Fort Collins, Colorado. 23rd-26th July 1985. General Technical Report A 13.88: INT-220. Ed. Robert C. Lucas. Ogden, UT: Intermountain Research Station, 1987: 246-293.

A review not only of modern research but of historical attitudes toward wilderness as these are reflected in major works on the subject. Although they must of necessity omit details and subtleties of opinion, the authors manage to summarize in a few succinct pages the shifting wilderness values of Americans since colonial times. The bulk of the article reports and interprets the results of studies of wilderness user motivations and the factors influencing visitor behavior in the wilderness. The text covers research and theoretical writings on the psychological nature of the wilderness experience as well as more conventional studies of attitudes and opinions. There is an extensive bibliography.

THE WILDERNESS EXPERIENCE

236   Abbey, Edward.   *Beyond the Wall: Essays from the
      Outside*.   New York: Holt, Rinehart and Winston,
      1984.
A collection of essays most of which appeared in earlier
versions in periodicals or in the wilderness photography
books of the 1970s.   Abbey's ability to deftly change tone
from outrage to humor to clean concentration on the
momentary details of landscape and physical experience
guarantees that his prose is always refreshing.   His quick
shifts allow him to say in spare, angry words the things
he believes need to be said about human degradation of the
natural world and human greed and selfishness without
lapsing into despair or misanthropy.   For many defenders
of wilderness his refusal to utter the usual politic
pieties will provide a welcome catharsis.

237   Abbey, Edward.   *Desert Solitare: A Season in the
      Wilderness*.   New York: McGraw-Hill, 1969.
A vivid record of the author's tenure in the desert
wilderness as a park ranger at the Arches National
Monument in Utah before the National Park Service "improved" the area with a paved road and other amenities for
tourists.   Abbey also recounts a trip down the Glen Canyon
of the Colorado before it disappeared permanently beneath
artificial Lake Powell.   Abbey is known for his angry
prose and his contempt for industrial tourism.   And that
is here; but there is also humor, good-natured recognition
of his own irascibility, love of literature, and respect
for the natural world.   For Abbey, the existence of
wilderness, whether experienced or not, is a talisman
against totalitarianism; and love of wilderness is an
expression of loyalty to the earth--"the earth that bore
us, the only home we shall ever know, the only paradise
we ever need."

238   Abbey, Edward.   *Down the River*.   New York: Dutton, 1982.
Recollections of rafting and kayaking down some of
America's still wild rivers interspersed with occasional
pieces of environmental journalism and literary criticism.

Abbey calls himself a toucher and a feeler not a thinker. What he gives the reader is not so much description or geology as the sensory immediacy of his own here and now experience of the natural world--the rank, tangy odor of a wild bear, the warmth of a cup of cowboy coffee held in cold hands on a river morning, the excited rush of adrenaline as a raft approaches a big rapid. The rest is humor, sometimes boisterous, sometimes self-deprecating, sharp satire and outraged anger at what Americans in the "pursuit of paper profits and plastic happiness" are doing to the beauty and mystery of free-running western rivers. "Every river I touch," he writes, "turns to heartbreak." Too often what is one year a wild river is lost the next in the dead waters of dams and irrigation projects. Abbey's anger and outrage are legendary, but they are also literary devices that transform into public speech the inexpressible pain of living in a society where "nothing is more vulnerable than the beautiful."

239   Abbey, Edward.   *The Journey Home: Some Words in Defense of the American West.*   New York: Dutton, 1977.
Essays exploring the connection between wilderness and civilization in American places from the emptiness of the western deserts to the alleys of Hoboken. This collection contains some of Abbey's most lyrical and literary prose. At one point he calls himself a prospector for revelation searching for a "blinding light illuminating everything." What he finds instead is no ulterior significance, only the sometimes heartbreaking beauty of nature continually surpassing itself and the love that it inspires. What he fights with his words is the mindless destruction of this wild beauty in the service of greed and a profligate way of life. His fear is for a future in which there is no refuge from tomorrow's technological totalitarianism. The existence of wild country is the guarantee of freedom: without wilderness the world becomes an "urbanized concentration camp." His summary judgment is simply this: "The idea of wilderness needs no defense. It only needs more defenders."

240   Alcock, John.   *Sonoran Desert Spring.*   Chicago, IL: University of Chicago Press, 1985.
Short essays on the coming of spring to the Sonoran Desert of Arizona organized by month from February to June. Within each month the essays focus on different species of desert plant or animal life. John Alcock's expertise as a zoologist enables him to speculate with authority about the evolutionary and ecological origin of the behaviors of creatures as different as peccaries and hairstreak butterflies. Respect for the plain dignity of this spartan environment and the "pleasing diversity" of nature infuses his quiet yet eloquent prose. In spite of differences in style, he expresses an outrage as great as Edward Abbey's or Wallace Stegner's at the human arrogance that would "deny the desert its due." Against the evolved design of the desert tortoise, Alcock sets the monumental canal of the Central Arizona Project, a concrete scalpel bisecting the state to bring water to Phoenix and the lawns of tract developments. Desert life is marvelously

equipped to cope with spareness; yet it is vulnerable to the unpredictable invasions of civilization in the form of idle target shooting, cactus rustling, or racing ORVs, a "trivial human invention whose only function is to amuse while destroying."

241   Alcock, John.   *Sonoran Desert Summer*.   Tucson, AZ: University of Arizona Press, 1990.
A continuation of the author's observations and speculations on the diversity of live in the Sonoran Desert of Arizona.   In this volume John Alcock begins where he left off in *Sonoran Desert Spring*--with the death of an ancient saguaro.   Now four years later in the drought of summer the tree's skeletal ribs lie like an "epitaph" on the desert floor.   Still, the desert organisms know the "seasons of the desert" and carry on the "delicate highwire act of organizing their lives" for their own well-being and for the success of their offspring.   Zone-tailed hawks mimic harmless turkey vultures, deceiving unsuspecting prey species, while Harris hawks hunt cooperatively to insure protein for individuals.   Biology, he cautions, does not have all the answers; and alternative explanations exist for most adaptive behaviors. Alcock's strength lies in his ability to evoke respect for the variety of patterns and solutions living things employ to solve evolutionary dilemmas and in demonstrating how biological thinking leads to the ecocentric insight that there is no "absolute scale of characteristics in the animal kingdom, running from primitive, inferior, and second rate to advanced, superior, and first rate."

242   Baron, Robert, and Elizabeth Darby Junkin, eds.   *Of Discovery and Destiny: An Anthology of American Writers and the American Land*. Golden, CO: Fulcrum, 1986.
A collection of writings organized in three major sec-tions: "Nature and Discovery;" "Nature and Challenge;" "Nature, Peace, and Self-Definition." Dipping into this book gives a sense of the variety of ways Americans have experienced the land.   The selections are brief excerpts but provide enough of a taste to lead interested readers to the original works.

243   Bergson, Frank, ed.   *The Wilderness Reader*.   New York: New American Library, 1980.
A judicious selection of wilderness writings illustrating the increasing concern over time, from William Byrd in the early 18th century to our contemporary writers, with preserving some remnant of the now almost unimaginable beauty and richness of the North American continent.   In his introduction, the editor calls attention to the theme of irretrievable loss, which appeared early and has grown darker and more urgent with the awareness that much of our vaunted technology is underlain with ignorance.   Recent writers must appreciate and describe wilderness against a background of unforeseen disasters resulting from the indiscriminate use of supposedly harmless materials and technologies.

244   Beston, Henry.   *The Outermost House.*   New York: Viking,
      1962.
      A reprinting of the 1928 book that set the standard for
      modern natural history writing.   Henry Beston records a
      year spent living alone in a cottage on the eastern arm
      of Cape Cod facing the Atlantic coast.   The overall
      organization of the book follows the seasons of the year,
      beginning in September and continuing through the follow-
      ing August; yet the author varies this seasonal pattern
      with chapters focused on special subjects, the changing
      voices of the waves, the mysterious beauty of the natural
      night or the neglected smells of sea and sand.   "A year
      indoors is a journey along a paper calendar; a year in
      outer nature is the accomplishment of a ritual."   Beston's
      language of color, sound and smell conveys the immediacy
      of his experience of nonhuman rhythms and reinforces his
      theme of the importance of wild nature unharassed by man
      to achieving dignity in human life.   In the foreword
      written in 1949 to accompany this edition he explains that
      without some awareness of the nature that is part of our
      humanness, "man becomes, as it were, a kind of cosmic
      outlaw, having neither the completeness and integrity of
      the animal nor the birthright of true humanity."

245   Brooks, Paul. *Roadless Area.* New York: Knopf, 1964.
      Descriptive, sometimes humorous accounts of wilderness
      camping trips the author and his wife made to national
      parks and other wild areas.   The selections are relatively
      brief, but Paul Brooks manages to capture the unique
      qualities and feeling tones of different ecosystems.   He
      also muses on the meaning of wilderness to people: "There
      is something refreshing to the soul in sharing your place
      in the sun with an animal that neither attacks you nor
      flees at the sight of you."   Opening wilderness areas and
      national parks to hunting destroys the possibility of
      experiencing this sense of being part of a world you do
      not control.   So does dependence on motorized travel; the
      Brookses canoe or walk and find that the very discomforts
      they endure are somehow subtly wound up with their
      appreciation of the integrity of place.   The book con-
      cludes with a chapter on the history of attitudes towards
      wilderness.

246   Carrighar, Sally.   *One Day at Beetle Rock.*   New York:
      Knopf, 1944.

247   ---.   *One Day at Teton Marsh.*   New York: Knopf, 1947.
      Descriptions of wild animals as they interact with each
      other and with their environment in the Teton Mountains
      and on Beetle Rock in Sequoia National Park.   The author
      imaginatively transforms her scientific knowledge and her
      careful observations so that the reader experiences the
      animal's lives and sensations from the inside.   She does
      not humanize them, but, instead, enables us to follow
      their inner impulses and instincts and to participate in
      their fears and expectations.   Although each chapter in
      these books emphasizes a different animal, events overlap;
      and the effect of the whole is to display the "willing
      tension which keeps a wilderness society stable."

248   Carson, Rachel.   *The Edge of the Sea.*   New York:
         Houghton Mifflin, 1955.
         The beautifully written story of intertidal life as it
         adjusts to the ebb and flow of the ocean on the rocky
         shores, sand flats and coral reefs of the Atlantic coast.
         The author finds in the adaptations of the plants and
         animals of the tidal zone a model of the toughness and
         tenacity of life itself. Rachel Carson conveys the daily
         existence of simple sea creatures with the exact knowledge
         of the scientist and the rhythmic prose of the poet. But,
         above all, she understands and expresses the value of the
         "intricate fabric of life by which one creature is linked
         with another and each with its surroundings."

249   Cook, Sam.   *Up North.*   Duluth, MN: Pfeifer-Hamilton,
         1986.
         Brief essays relating the author's adventures in the
         Quetico-Superior country. Sam Cook writes for the outdoor
         pages of the *Duluth News-Tribune & Herald*, where these
         pieces were originally published. They tell of hunting
         and fishing trips, winter camping and dogsledding, and
         canoeing ice-rimmed rivers in the northern spring. Cook
         is not a deeply philosophical or imaginative nature
         writer. His strength lies in his good-humored informality
         and the delight he takes in details of life on the trail
         and in the woods. Sometimes he sets out alone or with his
         dog to experience solitude in the wilderness. More often,
         he conveys his joy in visiting the north woods with
         friends and family. For Cook, watching the Northern
         Lights, listening to howling wolves or simply setting up
         camp are wilderness experiences that increase in value
         when they are shared.

250   Dillard, Annie.   *Pilgrim at Tinker Creek.*   New York:
         Harper's Magazine Press, 1974.
         Lively meditations exploring the exuberance and profusion
         of the natural world in language as intricate and vigorous
         as the events themselves. The author combines insouciant
         slang and contemporary colloquialisms with echoes of the
         Bible and extended fields of imagery--the magic show, the
         art gallery, the hunt--to convey the immediacy of her
         experience of the landscape and wildlife of the southern
         Virginia mountains. Her purpose is to learn to see--and
         to teach her readers to see--as a way of life. "Self-
         consciousness is the curse of the city." Alone in the
         semi-wild country of Tinker Creek Dillard finds it
         possible to concentrate consciousness in the pure devotion
         to the object that reveals the intricacy of creation. Her
         season in the wilderness is a pilgrimage that at unexpect-
         ed moments leads her to existential gaps through which she
         views the world transfigured. Dillard's spiritual
         adventure is a unique and provocative contribution to the
         genre of natural history.

251   Douglas, William O.   *Farewell to Texas: A Vanishing
         Wilderness.*   New York: McGraw-Hill, 1967.
         Visits to abused wildernesses in the state of Texas.
         Justice Douglas finds that, because no land in Texas
         originally belonged to the United States government,

little has been protected.   Big Bend National Park,
beautiful as it is, could be acquired by the federal
government only because ranchers had so overgrazed the
land that it had become worthless desert to them.   Rather
than see even the reduced remains of Big Thicket forest
set aside as wilderness, vandals, perhaps in the pay of
timber companies, attempt to destroy its scenic and
ecological value.   Douglas describes a thousand year old
magnolia tree that has been poisoned with arsenic to
prevent it from becoming a focus for preservation
efforts.   In the six years it took Douglas to complete the
field work for this book, he heard every one of his
outdoor values "appraised largely in terms of dollars."
The "modern Ahabs," as he calls them, are "more strongly
entrenched in Texas than anywhere else."   "That is why
this is a melancholy book."

252   Douglas, William O.   *My Wilderness: East to Katahdin.*
Garden City, NY: Doubleday, 1962.
Memories of times spent in the American wilderness from
the Rocky Mountains in the West to Maine and Mount
Katahdin in the East.   Justice Douglas visits these varied
areas, often in the company of the famous conservationists
associated with them--Olaus Murie in the Wind River
region, Sigurd Olson in the Quetico-Superior, and Harvey
Broome in the Smokies.   Yet each place becomes his own as
he reacts to local beauty and thinks about the meaning of
wilderness to the American people.   In contrast to the
claim that the wilderness is elitist, Douglas sees it as
the essence of democracy to preserve some parts of the
natural world as sanctuaries for our burgeoning urban
population.   The privileged few are not the hikers and
backpackers who visit the wilderness and leave no trace
but the ranchers and mining companies who, abetted by the
federal bureaucracy, are permitted to degrade and destroy
public lands.

253   Douglas, William O.   *My Wilderness: The Pacific West.*
Garden City, NY: Doubleday, 1960.
Recollections of the author's encounters with the western
wildernesses, particularly, the mountains and beaches of
the Pacific Northwest.   Justice Douglas grew up near the
Cascades; so the wilderness areas of Oregon and Washington
were a part of his life year after year.   His essays are
not so much records of particular experiences as
palimpsests of all his many trips, alone and with friends
and guides, over half a lifetime.   Anecdotes abound.
Remembering the pleasures of sleeping beneath a towering
sitka spruce with the sound of a wild beach in his ears
brings to mind another such a time when scavenging skunks
invaded the camp and marched heedlessly over a nervous Sig
Olson holding his breath in a tightly closed sleeping bag.
Old jokes are recalled, as are big trout caught in clear
streams; but Douglas never loses sight of his theme--the
importance of wilderness to the development of a complete
human being.   The words "cathedral," "sanctuary," and
"sacred" appear in almost every essay to reinforce his
message that, although a civilization can be built around

the machine, "it is doubtful that a meaningful life can be produced by it."

254   Ehrlich, Gretel. *The Solace of Open Places*. New York: Viking, 1985.
Not so much about wilderness as about the people who live around and near the still open country of Wyoming. The author does not discuss wilderness values, but her sense of the healing, almost spiritual quality of open spaces and her closeness to the natural world make these values apparent. Though she recognizes the negative effects of isolation from the city/suburbia of most of the United States on politics and sometimes even sanity, her love and respect for this spare land is evident in every word.

255   Eiseley, Loren. *The Immense Journey*. New York: Random House, 1957.
An imaginative and scientific representation of the unity not only of all life but of the process that brings life into being. In coming face to face with the skull of a primitive mammal as he explores a geological slit in the midwestern prairie, the author recognizes for a moment a "shabby little Paleocene rat," the "father of all mankind." He floats, giving his body to the flow of the meandering Platte River, and experiences himself as the momentary shape of the life that began in water. Peering out the window of a New York City hotel in a sleepless pre-dawn, he finds himself viewing the human skyscraper world from the inverted perspective of flocks of nesting pigeons. Eiseley calls these uncommon ways of looking at the world, in whatever environment they take place, "going into the wilderness." His essays serve as poetic reminders of the limits of a solely anthropocentric point of view.

256   Finch, Robert. *Common Ground: A Naturalist's Cape Cod*. Boston, MA: Godine, 1981.
Short, meditative essays exploring a personal coming to terms with nature as it exists in a particular time and place. For the author this place is Cape Cod as it reveals, even in its late century developed state, independent wilderness and nonhuman life lived on its own terms. Finch values wilderness as the entrance to a world apart from human standards that nonetheless calls to us in a language both "nourishing and irresistible." He writes of grackles on the lawn, the family life of a pair of local foxes, and deserted winter beaches. The least of things--ants in firewood or moths fluttering at the door--inspire speculations on the ways, dominating or anthropocentric or imaginative, that we try to relate to the natural world. Each essay begins with an actual happening described in concrete but evocative prose followed by a meditative questioning of his reactions and feelings, especially, the ambivalence he experiences toward what he calls the "hollow mind of the universe."

257   Finch, Robert. *The Primal Place*. New York: Norton, 1983.

Musings on life and death in Cape Cod. Though not a
wilderness area, the author's home in the hills beyond the
town and nearby tide flats of Cape Cod serve, in spite of
increasing development, as the site of his attempt to
enter the maze of natural life. As a human being, he
finds himself often an awkward outsider whose presence
disrupts the settled pattern of animal activity, one whose
ecological sentiments contrast strangely with his angry
bludgeoning of a garden-raiding woodchuck. Still, the
place possesses him, and he comes increasingly to recognize caring for others, both human and nonhuman, as an
inherent part of the human psyche. Finch discovers that,
paradoxically, the experience of being part of a whole one
loves comes when one is alone in the broad, spacious
setting of the natural world.

258   Frome, Michael. *Promised Land: Adventures and*
        Encounters in Wild America. New York: Morrow, 1985.
Vignettes of Americans enjoying the wilderness. The
author, a well-known conservationist and writer on
environmental topics, reminisces about his trips to the
national parks and wilderness areas of the United States
and the people he meets rafting or backpacking or pack-
horse camping. Frome knew some of the legendary wilderness advocates and recalls his experiences with Harvey
Broome and Sigurd Olson. But he also introduces the
reader to some less famous supporters of wilderness
preservation and to some otherwise ordinary people who
have chosen to sacrifice the comforts of urban living to
be close to nature. A salient wilderness value for Frome
is the equalizing effect of wilderness recreation. It
strips away distinctions of class and wealth so that
people are able to relate honestly to one another and to
themselves.

259   Furtman, Michael. *A Season for Wilderness: The Journal*
        of a Summer in Canoe Country. Minocqua, WI:
        NorthWord Press, 1989.
The contemplative diary of a summer spent as a Forest
Service volunteer ranger in the Boundary Waters Canoe Area
Wilderness. Michael Furtman writes in clear, uncluttered
prose of the beauty of the north country and of the deep
satisfaction he and his wife found cleaning and repairing
overused campsites, watching wildlife, assisting troubled
canoeists, and creating a simple life that left the
wilderness undamaged by their presence. He also writes
with anger and disgust at the abuse the land suffers from
the ignorance of campers, the difficulty of enforcing
regulations, and the failure of the Forest Service to
adopt quota systems sufficient to protect wilderness
values, particularly, the solitude and quiet that encourage visitors to become one with the rhythms and moods of
the land. But regulations, he feels, are not enough; a
new ethic that recognizes man as a part of nature and
respects the role of wild species in the ecosystem will
be needed before the selfish destruction and arrogant
unconcern he witnessed among certain campers will disap-
pear. Furtman is a long-time lover of the Quetico-
Superior country who delights in discovering the special

places sacred to earlier canoe country enthusiasts. His book joins theirs in chronicling the ongoing story of this unique area.

260   Gilbert, Bil.   *In God's Countries.*   Lincoln,   NB: University of Nebraska Press, 1984.
Witty, colloquial essays on wildlife and wild lands originally published in *Sports Illustrated* or *Audubon* magazine. Naturalist Bil Gilbert writes humorously but with respect about the resourcefulness of creatures as diverse as an idiosyncratic moose named the "Missouri Kid' and the nightcrawler tribe whose dedicated tunneling maintains the fertility of the soil in our gardens. His broad sympathies extend even to the entertainment hunters who rendezvous on opening day of deer season in the bars or along the roads of Potter County, Pennsylvania. The central theme that runs through these essays is the human yearning for contact with what Gilbert calls, after C. S. Lewis, "other bloods," the creatures that inhabit the earth along with us. This fascination with other animals is a "singular and definitive characteristic of our species." In the essay titled "Trailing" he shows this urge to understand translated into the excitement and satisfaction of the nonviolent sport of finding, following, and interpreting the tracks and other evidence of their presence that animals leave behind in the wilderness.

261   Gilbert, Bil.   *Our Nature.*   Lincoln, NB: University of Nebraska Press, 1986.
More essays on many topics, all illustrating what Bil Gilbert calls the old and honorable practice of nature loving. This is his straightforward term for the activities and attitudes that lie at the bottom of our efforts to preserve wilderness and wildlife. Rather than being an elitist avocation, going off into "surroundings which are relatively unaffected by man" is "close to being our most popular form of recreation, the commonest sort of esthetic activity and a very general source of satisfaction." Nature loving expresses itself in a dignified old man returning year after year to angle for trout in a country stream as well as in political efforts to protect the grizzly. As a nature lover himself, Gilbert finds contentment contemplating the slowly recovering ecosystem of the West Branch of the Susquehanna from a drifting canoe and adventure in snowshoe camping in nearby Potter County. This is not to say that he minimizes the importance or the appeal of large wilderness areas. His account of a canoe trip following the route of explorer John Franklin up the Yellowknife River from Great Slave Lake in the Northwest Territories accurately conveys the excitement and hardship of traveling in wild country and expresses his forthright appreciation for pristine natural beauty.

262   Graves, John.   *Goodbye to a River.*   New York: Knopf, 1961.
A canoe trip down the upper middle Brazos River in 1960 before a series of dams were to change its character

forever.   The author travels alone, except for the
companionship of a naive dachshund pup, along the narrow
strip of semi-wilderness that lies on either side of the
Brazos as it flows through the counties of Palo Pinto,
Parker, and Hood.   This is land John Graves describes as
cottoned out in the flat places and grazed out on the
slopes in the 19th century by the settlers who destroyed
and replaced the prairie-roving Comanches.   Graves has
stories to tell of pioneer days in places along the river;
but the motive for his trip is less historical than
personal--to experience the paradoxical solitude of the
wilderness, in which being alone entails being part of the
natural whole, intensively alive to sights and sounds and
weather.

263   Gruchow, Paul.   *Journal of a Prairie Year*.   Minneapolis,
           MN: University of Minnesota Press, 1985.
       A book of the seasons as they pass in the remnants of the
       old tallgrass prairie of the Middle West.   Throughout, the
       author contrasts the surprise, profusion and complexity
       of life on the wild prairie with the geometrical,
       straight-edged tendency of industrial agriculture and
       commercial development.   In each of his descriptions
       Gruchow strives to see and, so, in a sense, to enter and
       possess the wild world.   He relates that, as a youngster,
       he tried to learn the life of the mink through trapping,
       until he realized that such learning was purchased by
       destruction of the life he admired.   Instead he became a
       watcher and was rewarded with the glimpses of the secret
       life of the land that he shares here with the reader.

264   Gruchow, Paul.   *The Necessity of Empty Places*.   New
           York: St. Martin's, 1988.
       A record of walking westward and upward from the farmlands
       of southern Minnesota and the Nebraskan prairie into the
       austerity and silence of the Big Horn Mountains.   In
       between, the author's metaphorical eye functions telescop-
       ically to reveal landscapes of vast distance and spare
       heights or shifts to microscopically examine the lillipu-
       tian plant world of a fell-field.   As he walks and camps
       in the mountain wilderness, Gruchow meditates on the
       values of empty places--the spiritual joy that counter-
       points deprivation, discomfort or, even, danger on the
       trail and the authentic experience of place that awards
       concentration in solitude.   He notices, too, and mourns
       the deficiencies of modern America--the tendency to
       homogenize the environment and the childish arrogance of
       our belief that we have conquered nature.   The pervading
       theme of these essays is the continuing human longing to
       escape the alienation of civilization and find acceptance
       in the unfathomable community of the wild.   Paul Gruchow
       is an interesting new writer of natural history.

265   Halpern, David, ed.   *On Nature; Nature, Landscape, and
           Natural History*.   San Francisco, CA: North Point
           Press, 1987.
       An intriguing anthology that stretches the genre of nature
       writing to encompass history, philosophy, biography and
       fantasy.   The collection opens with Noel Perrin's analysis

of the 1950s as the "swing point in man's relationship to nature," when we became capable of wrecking the planet, and ends with John Rodman's "The Dolphin Papers," a science fiction interpretation of human civilization from the cetacean point of view. In between, a variety of writers reveal their experience of the natural world. John Fowles negates the conventional opposition of nature and art by contrasting the creative energy and indeterminacy of both with our numbing addiction to labeling and controlling. Barry Lopez and John Haines express in image and symbol the necessity to the human spirit of preserving our evolutionary and archaeological past. Lopez's intuition that their destruction creates the "awful atmosphere of loose ends in which totalitarianism thrives" rings true. In "Shadows and Vistas," a reprint of the keynote address given at the Alaska Environmental Assembly in 1982, Haines sees the result of the destruction of wilderness and wild things as poverty, "an inner desolation to match the desolation without."

266 Hay, John. *Immortal Wilderness*. New York: Norton, 1987.
Reflections on the theme of human alienation from wilderness. John Hay enlarges the definition of wilderness to encompass nothing less than universal interdependence and the foundation of all life. What we usually call wildernesses are the remnant areas in which natural processes of renewal and diversity still predominate. In our self-righteous rush to assault and occupy the continent and the planet as human domain, we make nonpersons out of the earth's other inhabitants. Ironically, humans are the ultimate losers, imprisoned in a world that "only looks out the window at itself." Hay uses his own experiences of nature as well as evocative imagery of space and light and sky to shatter the rigidity of his readers' minds and permit a glimpse of the immense wilderness beyond the self, where each community expresses the "profound capacities of life in its own way." These short essays serve as stimulating antidotes to the "deforestation of the spirit" the author deplores.

267 Hay, John. *In Defense of Nature*. Boston, MA: Little, Brown, 1969.
Passionate meditations on technological man's brutal dominance of nature. John Hay uses the intricacies of plant and animal life in the water, sky and woods near Cape Cod to show that the exploitative power that insulates humans from the natural world also isolates us from reality. We dismiss the close fittingness of the heron's fishing technique and the remarkable attunement of all its senses to its environment as limited behavior and put down the universal "rightness of life and place" with arrogance born of our material systems. Yet a tree is not just so many feet of board timber or merely an ornament for our gardens but a live environment for native birds and squirrels and the thousands of microorganisms that thread their ways through roots and leaf mold. The dovekies dashing and diving off Cape Cod nest on drift ice in the artic, where they support life for predators--gyrfalcons

and arctic foxes--in an otherwise barren land. The truth
of existence is this interdependence from which, in spite
of our "death-dealing capacity," we are not excluded.
"Man against the natural world is man against himself."

268   Hay, John.   *The Undiscovered Country*.   New York:
        Norton, 1981.
        Poetic, sensitive renderings of the author's attempts to
        participate in the web of life that surrounds his home on
        Cape Cod. John Hay praises the common animals and plants
        of his chosen space as they respond to the ancient rhythms
        of their existence, migrating vast distances, storing food
        for the local winter or competing in spiraling spring
        song. He marvels at the seemingly miraculous response of
        birds and fish to changing atmospheric conditions and
        envies the box turtle's unerring sense of place.   The
        phrase "homing in" becomes a metaphor for his recurring
        theme of what humanity has lost in separating itself from
        the natural world. We deceive ourselves when we think we
        can live on rather than with the land. Hay calls us to
        wake up from our somnambulant human world to the height-
        ened reality of sharing in all of life. Our true home is
        open to the beauty and terror of existence.   "In our
        lasting domesticity are wild directions."

269   Hoagland, Edward.   *Red Wolves and Black Bears*.   New
        York: Random House, 1976.
        Intensely personal essays connecting New York City to the
        wilderness and dogs to wolves through the author's
        conviction that life at its evolutionary basis is good
        whether for a wolf or a dog or a man. Edward Hoagland
        writes with verve, affection and admiration about our
        remaining big predators--bears, wolves and mountain
        lions--and about predator people--the assorted wildlife
        biologists, trappers and outdoorsmen who make knowing
        these animals their life work. He captures the special
        zest for life of animals in energetic prose: wolves and
        dogs "grin and grimace and scrawl graffiti with their
        piss," and a young male bear released from a trap bounds
        toward the woods "like the beast of a children's fairy
        tale--a big rolling derriere, a big tongue for eating, and
        pounding feet, its body bending like a boomerang." Yet
        mixed with the joy he conveys is sadness that so much of
        what is wild is already gone and we are "hardly aware of
        what remains." Wilderness itself is indifferent even to
        its own destruction; we are the ones who impoverish
        ourselves by destroying the rich tapestry of the world.

270   Hoagland, Edward.   *Walking the Dead Diamond River*.   New
        York: Random House, 1973.
        A mixture of pieces on various topics, including two
        essays on the wilderness: "The New England Wilderness" and
        "Walking the Dead Diamond River." Both essays, though
        filled with landscape detail and curiosity about wild
        animals and plants, conclude that there is "no wilderness
        as such left in New England." The problem is that any
        officially designated wilderness is overrun with visitors
        so fast that it ceases to provide a wilderness experience,
        while any de facto wildernesses are being rapidly sold off

for development. Hoagland also finds that most people are
losing their love for the wilderness and no longer really
want to roam the woods on their own.

271   Hoover, Helen. *A Place in the Woods*. New York: Knopf,
         1969.
The enthralling story of the author's precipitous change
in life style in 1954 from an engineering career and
apartment living in Chicago to freelance writing in a
cabin in the northwoods. Though the Hoovers start their
adventure with limited savings and no regular source of
income, they refuse to give up the way of life they have
chosen even as they cope with one disaster after another.
Ade Hoover is handy with tools, and Helen has a strong
scientific background; but neither of them is an experienced outdoorsperson. They find they must contend in
their first year with a collapsing foundation, leaking
roof, fire and loss of their automobile, all in rain or
snow or subzero temperatures. Living in the wilderness,
they learn not only to accept but welcome the unexpected-
ness and variability of nature "outside of human planning." They are sustained by the beauty of their surroundings and by their empathy with wildlife and openness
to new experience.

272   Hoover, Helen. *The Years of the Forest*. New York:
         Knopf, 1973.
The story of thirteen years from 1956 to 1971 living on
a border lake in the Quetico-Superior, learning and loving
and protecting the wild creatures that share their forest
home. Helen Hoover and her husband live ecocentric lives,
adapting their activities to the requirements of the land
and the needs of wild species. They do not think they are
"any more important than the flora and fauna" around them.
When a wandering woodchuck moves in, they build a fence
around their garden instead of shooting him; and, when
they are invited to install electricity, they decline
because it would entail cutting down their ancient pines.
They are rewarded by the trust of their animal friends but
ostracized by unsympathetic locals who call them "kooks"
and "crackpots" for protesting DDT spraying and defending
deer on their property against trespassing hunters.   In
the early years alone in the woods they are poor and
sometimes hungry. Later, after Helen's books have become
a financial success, they can afford a few comforts.
Ironically, this is also a time when electric power, paved
roads, summer homes and the activities of commercialized
resorts bring about a gradual attrition of wilderness.
Still, they stay; for even a diminished wilderness has the
"magic to heal the wounds made by men."

273   Hubbell, Sue. *A Country Year: Living the Questions*.
         New York: Random House, 1986.
Everyday adventures of a woman who leaves a secure middle-
class life for the "wild, anarchistic, raffish appeal" of
beekeeping on a hundred odd acres in the Ozarks. Sue
Hubbell's deceptively simple and straightforward essays
produce a cumulative effect as they gradually reveal the
wisdom she has gained living with the land. Her theme of

letting go of control and dominance to take one's place
in the interdependent circle of life is subtly introduced
in the first essay where she whimsically mulls over the
claims of various nonhuman residents to her property.   In
later essays she experiences herself from a bat's point
of view as bait to attract mosquitoes and participates in
the elation of a spring swarming of bees, although for
them she is merely an object to be encircled.   Humans are
natural meddlers and fiddlers; a chainsaw makes a difference in the woods.   But humans can also understand the
reverberations of their actions.   Hubbell discovers that
it is possible to live productively without obsessively
attempting to control the life around her.   A healthy
colony of bees will take care of its own cockroaches, and
an unmown meadow holds surprises of grazing deer for an
eye that has "long ceased to find trimness pleasing."

274   Kesselheim, Alan S.   *Water and Sky: Reflections of a*
        Northern Year.   Golden, CO: Fulcrum, 1989.
The enthralling narrative of a canoe trip through the
Canadian wilderness.   Alan Kesselheim and his partner
Marypat Zitzer begin their two thousand mile water journey
in June on the Athabasca River at Jasper, Alberta and
finish over a year later at Baker Lake just below the
arctic circle in the Northwest Territories.   In between,
they winter over in a log cabin at a summer fishing camp
for tourists on Lake Athabasca.   Physical challenge
becomes part of their everyday life as they tackle tricky
rapids, high winds, threatening waves, voracious insects
and a marauding bear.   They are rewarded with the pristine
beauty of truly wild country, the tundra barrens and the
rock-walled lakes, and sightings of caribou, musk oxen and
wolves carrying out their ecological roles undisturbed by
humans.   In their life style of "profound simplicity" they
experience wilderness as an "empowering serenity."   There
are times, though, of poignant sadness.   They mourn the
death of a member of a familiar wolf band who has been
heedlessly shot, and regret the passing of the ways of the
native elders in the plastic squalor and mechanized racket
of modern settlements.   Finally, they must face the
dislocation and wrenching pain of leaving the wild country
to rejoin civilization.

275   Krutch, Joseph Wood.   *The Best Nature Writing of Joseph*
        Wood Krutch.   New York: Morrow, 1969.
A collection of essays selected by the author from his
writings on the animal and plant life of the Connecticut
countryside, the Sonoran Desert and the Baja peninsula.
The essays combine a clear, flexible prose style, wit and
literary allusion with careful observation and scientific
awareness to express the author's joy in being part of the
community of living things.   All the strains of wilderness
appreciation appear, with varying emphasis, threading
their way through the collection.   Nevertheless, underlying and serving as a counterpoint to his dominant message
of human delight in the intricacies of the natural world,
is the gnawing fear that humankind, in an arrogance of
greed and growth, will bring about a totally artificial
and mechanical civilization, in which all joy and freedom

are crushed by the hard necessity of constantly monitoring
the earth to prevent the natural disasters made inevitable
by the destruction of natural ecosystems.

276   Krutch, Joseph Wood. *The Desert Year*. New York: Sloane,
       1952.
       A record of the author's first year in the Sonoran Desert,
       while on leave from his teaching duties at Columbia
       University.   Krutch spends his year in this place of
       little rainfall, so different from the New England
       suburbs, learning to see and hear the natural world.   In
       the end he discovers what he set out to find--joy in the
       awareness of the marvelous ingenuity and courage of wild
       animals and plants as they endure and thrive apart from
       human control or intervention.   He also reaffirms some
       forgotten platitudes about the precariousness of life and
       the meaning of abundance.

277   Krutch,   Joseph   Wood.   *The   Forgotten   Peninsula: A
       Naturalist in Baja California.   New York: Sloane,
       1961.
       Vignettes of Baja in the 1950s when most of the peninsula
       was de facto wilderness.   Joseph Wood Krutch relates his
       own adventures and muses on the vegetative strangeness of
       Boojum trees and the correlation between bad roads and
       beauty with gentle humor, companionability and love of the
       land.   He ponders the price Baja will pay for progress,
       not only aesthetically but also in the quality of the
       lives of its people, who may find they have traded
       laughter and leisure for possessions.   Nevertheless, he
       is not so naive as to believe the Baja he visited to be
       a pristine paradise.   Already by the 1950s most of the
       larger mammals, both predators and prey, had been elimi-
       nated.   Man, wherever he exists, is the "only animal who
       habitually exhausts or exterminates what he has learned
       to exploit."

278   Krutch,   Joseph   Wood.   *The   Voice   of   the   Desert: A
       Naturalist's Interpretation.   New York: Sloane,
       1955.
       A sensitive tribute to the Sonoran Desert and to the
       ingenuity and endurance of its plant and animal inhabit-
       ants.   Each essay records Joseph Wood Krutch's observation
       of a desert species--the spadefoot toad, the road runner,
       the saguaro cactus, the yucca moth and others--and
       considers its adaptations to the arid Arizona climate.
       Yet each goes beyond biological science to what Krutch
       calls "metabiology," rational but sometimes heretical
       speculation regarding the role of courage and choice in
       evolution.   The concluding essay, "The Mystique of the
       Desert," is an emotionally powerful defense of the desert
       (and all wilderness) as the landscape of contemplation and
       source of the mystical experience of being part of a
       greater whole.   This collection also contains Krutch's
       prescient and often reprinted essay "Conservation is Not
       Enough," which years after its publication remains a
       timely and eloquent statement of the need for an ecocen-
       tric philosophy.   The enlightened selfishness of resource
       conservation can only "defeat itself and the earth will

have been plundered no matter how scientifically or
farseeingly the plundering has been done."

279   La Bastille, Anne.   *Beyond Black Bear Lake*.   New York:
        Norton, 1987.
      The continuing story of a woman alone on thirty acres in
      the Adirondack wilderness attempting to live in harmony
      and community with native plants and animals.   In this
      sequel to *Woodswoman*, ecologist Anne La Bastille recounts
      building a simple retreat--she calls it Walden II--on
      Lilypad Lake to get away from the encroachment of civili-
      zation on Black Bear Lake in the form of highpowered
      motorboats and intrusive sightseers.   She shares with her
      readers not only her love for the wild country of the
      Adirondacks but also her personal feelings and fears.   She
      confides her overwhelming sadness at the death of her
      German Shepherd Pritzi, her slowly awakening emotional
      commitment to Doctor Mike, and her dread of the physical
      disabilities of age that might force her to leave her
      wilderness home.   This is a darker book than *Woodswoman*.
      Even at Lilypad on her own land with no access but a
      narrow foot trail there is no escape from the effects of
      industry and exploitation.   In the heart of the wilderness
      acid rain has turned once productive lakes into lifeless
      pools, and the fish and the frogs and the otters have
      gone.

280   La Bastille, Anne.   *Woodswoman*.   New York: Dutton, 1976.
      The encouraging story of a single woman who seeks an
      advanced degree in wildlife ecology and practices as an
      ecological consultant, while living alone in the Adiron-
      dack wilderness.   Left without family and unable to
      continue her occupation as a lodgekeeper after her divorce
      from her husband, Anne LaBastille moves to a secluded
      piece of wild and wooded land accessible only by water in
      the summer and ice in the winter.   Building her primitive
      cabin is hard work; but, with help from friends, she
      succeeds in this and goes on to acquire the skills and
      strength necessary for taking care of herself in the
      wilderness.   In fact, she passes the outdoor tests for
      becoming an Adirondacks guide.   With her dog Pritzi as a
      companion, she overcomes occasional bouts of fear and
      loneliness.   Touching the ancient pines that surround her
      cabin, she receives a strange energy of communion.
      Watching wild animals becomes a source of delight as well
      as a means to scientific knowledge.   Extended professional
      trips to cities only confirm her preference for a simple
      way of life with minimal impact on the environment.   Yet
      she recognizes her dependence on the outside world.   The
      Adirondacks are too cold and too impoverished of fish and
      game to yield an independent living.   What they give
      instead is "amber wildness blazing up" in a fox's eyes and
      "ebony water" like a "mirror under yellow cow lilies."

281   Lehmberg, Paul.   *In the Strong Woods: A Season Alone in
        North Country*.   New York: St. Martin's, 1980.
      The unassuming record of a search for a way to live
      modestly and with contentment in both the natural and the
      human worlds.   Paul Lehmberg leaves the city life that

"pummels the senses" to live alone, with only his dog Zip for companionship, from May through September in a handbuilt cabin on Nym Lake in the Quetico country. In this glaciated land of rock, water and sky he discovers the psychological and aesthetic values of simplicity and strives to maintain them against the human tendency to collect physical and mental clutter. He learns to find satisfaction and clarity of mind not only in the immensity of the northern sky but in the repetition of domestic tasks--baking bread and sawing wood--with care and concentration. Towards the end of his stay, having become almost too comfortable in the hermetic life, he affirms the human need for community as well as solitude. His well-crafted book is, in part, an attempt to join these two worlds by sharing his wilderness experience with others.

282　Leopold, Aldo. *Round River*. Ed. Luna B. Leopold. New York: Oxford University Press, 1953.
Excerpts from the journals of Aldo Leopold edited by his son. The pieces are arranged chronologically to record the development in Leopold's perception of the ethical and aesthetic values to be found in natural things. The Leopold of 1922, hunting geese in the Southwest, automatically shoots to kill any coyote, mountain lion or raccoon that crosses his path. While camping, he casually baits traps for these animals without compunction or even thought. Years later, this same man will write eloquently that a conservationist cannot love game and hate predators, and ask the question, "Is a wolfless north woods any north woods at all?" He has learned to perceive the wilderness as a biotic community and found an ethical underpinning for relationship with the land.

283　Leopold, Aldo. *A Sand County Almanac and Sketches Here and There*. New York: Oxford University Press, 1949.
The seminal book in the development of environmental philosophy and the sacred text of the wilderness movement. This superb work examines man's relationship to the natural world in three sections, building from the particular and personal to the general and abstract. The first section, entitled "A Sand County Almanac," is organized around the months of the year at the Leopold family's worn-out farm in the sand counties of Wisconsin, where he spent his vacations and weekends working to restore the health of the land. Here the ecological connectedness of things is conveyed through everyday experiences of local animals and plants. The focus broadens geographically in "Sketches Here and There" to include all of Wisconsin and the United States, and the concern with ecological diversity and stability becomes more explicit. Nevertheless, these essays remain rooted in concrete personal experiences. The last section, called "The Upshot," is more philosophical, though Leopold's craftsmanship as a writer assures that each word is carefully selected and each sentence finely tuned. This section is the source of his famous essay, "The Land Ethic," in which he envisions human beings relating to the land ethically as members of the ecological community.

284   Lopez, Barry.   *Arctic Dreams: Imagination and Desire in
         a Northern Landscape.* New York: Scribner's, 1986.
      A moving evocation of the beauty and mystery of the Arctic
      landscape and its wildlife.   The author uses devices of
      poetry--cadence, alliteration and assonance--to convey the
      strangeness of a land lambent with light, light reflected
      from snow and snow geese and refracted in icebergs,
      contrasted with dark green or black water and showering
      from stars in the long winter night.   The respect for
      nature that environmental philosophers theorize becomes
      actual in Lopez' prose.   In the Arctic he discovers the
      innate dignity and integrity of the land with its plants
      and animals.   If these have no dignity, neither has
      humankind.   Our vaunted dignity becomes merely a shaky
      cultural invention, a kind of whistling in the dark, not
      a perception about reality.

285   Lopez,   Barry.   *Crossing   Open   Ground.*   New   York:
         Scribner's, 1988.
      A collection of essays originally published in periodicals
      during the 1980s.   Whether the subject is migrating snow
      geese at Tule Lake, rafting down the Grand Canyon,
      canoeing the Charley River in Alaska, or the dying of
      beached sperm whales on the Oregon coast, Lopez translates
      the mystery and beauty of nature and the human imagination
      into disciplined prose.   In the essay entitled "Landscape
      and Narrative" he meditates on a strangely resonant
      intuition of the value of wilderness.   The interior
      landscape of the mind--the ordered set of relationships
      of ideas, speculations, and feelings--responds to the
      "character   and   subtlety of   an   external   landscape."
      Wilderness is, thus, intimately related to our ethical and
      psychological well-being.

286   Lueders, Edward.   *The Clam Lake Papers: A Winter in the
         North Woods.* New York: Harper & Row, 1977.
      Meditations on language and life inspired by a solitary
      winter spent at the author's lakeside cabin in the
      Wisconsin woods.   Edward Lueders presents his winter
      experiences and thoughts as the comments of an unknown
      intruder left on his writing table for him to read in the
      spring.   This mysterious alter ego finds that solitude in
      the season of snow and cold enlarges his notion of society
      to include the sounds of inanimate things as well as the
      discourses of ravens and bluejays.   The present reality
      of Clam Lake isolation overwhelms the conventional now of
      current events on television. Sometimes thought runs down
      or runs out replaced by an "unfathomable richness of
      silence."   Occasionally, he experiences himself as an
      event in the consciousness of one of the animals who
      shares the woods with him.   One evening rapt in his
      writing he discovers that he has been all along the object
      of intent study by a curious fox peering through the
      lighted window. Lueders' book gives only minimal informa-
      tion about the north woods or even about daily life in the
      wilderness.   It explores instead the introspections of a
      literary mind alone in wild place.

287 Lyon, Thomas J., ed. *This Incomperable Lande: A Book of American Nature Writing.* Boston, MA: Houghton Mifflin, 1989.
A selection and analysis of natural history writings designed to provide the basis for a re-evaluation and a revisionist interpretation of the significance of the nature essay in American literature. Thomas Lyon structures the work in three sections: an analytical history of nature writing in America; a collection of chronologically arranged essays from William Wood in the early 17th century to modern writers of the 1970s and 1980s; and an extensive annotated bibliography covering secondary literature--philosophy, history and cultural interpretations--as well as first-hand responses to the natural world. Lyon sees the writers of natural history as a heretical, nonconforming minority who refocus our vision outward from the narrow egotistical expansionism and impoverished one-at-a-time perception of the dominant frontier mentality. The typical nature essay leads the reader to perceive the world as an ecological whole that ultimately includes the human observer, thus encouraging an ethical rather than an exploitative attitude towards wilderness and wildlife. This volume is not only a fine collection but a thoughtful examination of the characteristics and cultural role of a neglected genre.

288 Madson, John. *Where the Sky Began: Land of the Tallgrass Prairie.* San Francisco, CA: Sierra Club Books, 1982.
A natural history of the tallgrass prairie that once covered much of Illinois, Iowa, Kansas and Missouri, as well as western Minnesota and the eastern portions of the Dakotas. John Madson, a native Iowan whose grandfather was one of the pioneers who broke the prairie, writes with easy familiarity and scientific accuracy of the evolutionary history of the prairie wilderness and the ecology of its native grasses and flowers as they exist in remnant reserves even today. Quoting the wonder of early explorers and relating his own experiences hunting down vestiges of this unique American biome, he is able to convey something of a "light-filled wilderness of sky and grass" that appears uniform only to the uneducated eye. There is no mistaking the "indefinable shaggy, fierce look" of native prairie for the plowed fields and overgrazed bluegrass pastures that surround it. The difference between the "wild original and its spiritless descendants" is captured throughout Madson's book in language that contrasts wild and tame, stolid and lively or fixed and free. The final section discusses efforts to establish a prairie national park, gives directions for planting prairie grasses and lists the existing designated prairie reserves.

289 Marshall, Robert. *Arctic Wilderness.* Berkeley, CA: University of California Press, 1956.
The journals of Robert Marshall's expeditions into the Arctic wilderness and the Brooks Range throughout the 1930's, edited by his brother and published posthumously. The author's enthusiasm for wilderness exploration comes

through in his attention to the details of the land and
life in camp. Occasionally, his happiness is expressed
directly as appreciation both for the experience of the
immensity and grandeur of nature and for the sense of
peace and security he finds in being alone with only a few
devoted comrades in a land empty of human settlement.
Marshall realizes that there could never be enough
wilderness to afford the world's millions the experience
of discovery he had; yet he believes that there will
always be a minority for whom pristine nature is a basic
source of happiness, and he advocates wilderness preservation so that future generations can experience the
profound joy he derived from these trips.

290   McPhee, John.   *Coming into the Country.*   New York:
        Farrar, Straus and Giroux, 1977.
A look at Alaska from three perspectives: a canoe and
kayak trip down the Salmon River in the Brooks Range;
interviews with urban Alaskans about building a new
capitol; and wilderness outings and conversations with
frontierspeople living in and around the town of Eagle.
The contrast between the experience of the wild river in
the first section and the pioneering values expressed by
the settlers poignantly reveals a perennial problem
inherent in wilderness preservation of deciding at what
point those who wish to live in and off the wild country
begin to destroy it.

291   Milton, John P.   *Nameless Valleys, Shining Mountains:*
        The Record of an Expedition into the Vanishing
        Wilderness of Alaska's Brooks Range.   New York:
        Walker, 1970.
The journal of a 300 mile backpacking expedition to the
Alaskan wilderness in 1967.   The author walked with two
companions from the southern foothills of the Brooks Range
to the Barter Island DEW Line station on the Arctic Ocean.
Milton shares with Aldo Leopold and John Madson a delight
in the beauty of humble plants in their ecological
setting, but he is equally adept at recording the grandeur
of gray rock and volcanic red sunsets.   He believes that
wilderness is necessary for modern people to escape
absorption in the increasingly mechanized world and
relearn the origins of the human spirit in the natural
world.   But, ultimately, wilderness should be preserved
for its own sake, not ours.

292   Murie, Adolph.   *A Naturalist in Alaska.*   New York:
        Devin-Adair, 1963.
Observations of Alaskan wildlife in Mount McKinley
National Park during the 1930s, 1940s, and 1950s by a
biologist with the National Park Service, the brother of
Olaus Murie.   The author relates behaviors of the wild
animals he studied in spare, easy prose, frequently
referring to precise dates and times of his sightings;
nevertheless, this is an engaging and loving book.   Adolph
Murie combines his scientific observations with an
empathetic sense of the daily adventures of grizzlies,
wolves, foxes, Dall sheep, and even the humble haymaking
Toklat vole. The reader comes away convinced that, though

the wilderness is, in no sense a peaceable kingdom, it is a place where the inhabitants, unthreatened by man, enjoy the hunting, food gathering, and social communication that make up their lives. For Murie, the wilderness spirit of Alaska is just this free, bold existence of its wildlife.

293 Murie, Margaret. *Two in the Far North.* 2nd ed. Anchorage, Alaska: Alaska Northwest Publishing Company, 1978.
An enthralling narrative of life in Alaska, beginning in 1911, when the author was eight years old, and closing with a speaking tour in 1975 on behalf of the wilderness. In between, the reader learns what it was like to be a child in Fairbanks when the only contact with the outside world was by river steamer in summer and horsedrawn mail sledge in winter and shares Margaret Murie's happiness in her unconventional honeymoon traveling from Bettles to Wiseman by dogsled in 1921. Other equally adventurous trips follow. In 1926, she and her husband, Olaus Murie, take their infant son along on a bird banding trip up the Crow and Porcupine rivers; in 1956 they join a group of scientists in studying the area of the Brooks Range for designation as a wildlife refuge. Among the many books on Alaska, this is outstanding for Murie's complete and understanding portrayal both of the beauty and variety of the land and of the characteristics of the people for whom it is home.

294 Murie, Margaret, and Olaus Murie. *The Wapiti Wilderness.* New York: Knopf, 1966.
Heartwarming stories of the Muries' life in the Wyoming wilderness, beginning in 1927 with Olaus' assignment to study the Jackson Hole elk herd and continuing through the war years to the 1950s and his term as Director of the Wilderness Society. Separate chapters by Margaret and Olaus Murie alternate to evoke different aspects of their life. Simple details of camp and householding chores are interwoven with summaries of scientific and conservation work, including accounts of the recurring community conflicts over protecting the herd and extending the park. Occasional misadventures, whether serious or humorous, serve to highlight the harmonious family life the Muries build in their idyllic setting. The greatest pleasure of parents and children alike lies in watching the wildlife that share their world. They find enduring satisfaction in understanding and interpreting what they see. Wilderness appreciation appears not in seeking thrills but in the "small adventures occurring from day to day."

295 Murie, Olaus J. *Journeys to the Far North.* Palo Alto, CA: The Wilderness Society and American West Publishing Company, 1973.
A looking back from the vantage point of middle age on a lifetime of wilderness adventure and scientific investigation in the Alaskan and Canadian north. Olaus Murie begins his reminiscences with his canoe trip up the Great Whale River in Hudson Bay country as a young man just out of college in 1914 and closes with a journey with his wife Mardy to study the flora and fauna of the Brooks Range in

1956. A final section of the book ponders the meaning of the Arctic for man. Murie's direct, concrete prose style is perfectly suited to his artist's eye descriptions of the simple contrasting patterns of color and form in the northern landscape and to his understated accounts of the hardships and danger that accompanied travel by canoe and dogsled when much of the north remained only minimally explored. Among Crees and Eskimos his enthusiasm, openness and humility evoke friendship and trust; and his book is a rich source of information and anecdote about their cultures. Murie hunts wild animals to feed himself and his dogs and to prepare specimens for the agencies that employ him; yet his attitude toward wildlife is one of respect and appreciation. In saying farewell to the nesting birds of the tundra, he hopes "every spring the tundra ponds are graced by these little folk, for the fulfillment of their own destiny and the enrichment of man's spirit." For Murie science and aesthetic appreciation are not ultimately opposites. At the heart of scientific understanding there is "wonder and a degree of respect for what we only partially understand."

296 Nelson, Richard. *The Island Within*. San Francisco, CA: North Point Press, 1989.
A celebration of the misty, storm-besieged islands off the Alaskan coast and a progress report on learning to live with restraint as a plain member of the earth community. For anthropologist Richard Nelson this entails training his senses and imagination in the ways of the Koyukon elders. In his eagerly awaited journeys to his chosen island he disciplines himself to experience the natural world directly through his senses, resisting the tendency of modern scholars to rely on borrowed abstractions or confine felt experience in scientific data. He practices merging his identity with ancient trees and native animals, soaring and diving in a bald eagle's flight or quivering in an alarmed deer. His respect for nature evidences itself in the patience with which he learns to wait for a hunted deer to give itself and in the meticulous ritual with which he carves its body. Toward the end of the book he recounts at last a kind of epiphany in which he sees himself and all things as "nothing more nor less than the earth expressing itself in living form." This is a magically beautiful book full of the enchantment of impeccable devotion to true experience and exact language.

297 Olson, Sigurd. *Listening Point*. New York: Knopf, 1958.
Essays organized around the image of the wilderness as a listening point, a place where, if one comes in humility and openness, it is possible to experience wonder and beauty and sense of belonging to the universe. Olson's own listening point is a campsite on a rocky ledge covered with bearberry, culminating in a single wind-swept pine, looking out over a canoe country lake and, metaphorically, over the extent of the northern wilderness. Each individual essay, whether the subject is forest succession or canoe portages or glacial striae, begins and ends at Listening Point, echoing the overall organization of the

collection. This is the most aesthetically satisfying of the author's many recollections of the Quetico-Superior country.

298   Olson, Sigurd. *The Singing Wilderness*. New York: Knopf, 1958.
Essays that evoke the Quetico-Superior wilderness through-out the seasons of the year. The author's conviction that the experience of wilderness is necessary to living a harmonious human life is implicit in the detail of color, sound and smell he uses to describe the Northern Minnesota landscape and wildlife. It becomes explicit in essays like "Farewell to Saganaga," where he relates his disap-pointment and sadness in returning to this perfect wilderness lake after it was opened to road traffic. Instead of the silent singing of the wilderness, his ears are besieged by the mechanical shrieks and drone of an outboard motor and the strains of recorded dance music from the newly opened lodge.

299   Olson, Sigurd F. *The Lonely Land*. New York: Knopf, 1961.
The record of a canoe trip down the Churchill River in Saskatchewan from *Lac-Ile-a-la-Crosse* to Lake Cumberland and the Hudson Bay Company's Cumberland House on the route of the voyageurs. Acting as "Bourgeois," the leader of a voyageur brigade, Sigurd Olson travels with five companions, all expert canoeists, to relive the adventures of fur company explorers. Each evening they read aloud from the diaries of Mackenzie and others who wrote about this wild country when it was first seen by Europeans; everyday they run rapids tested and named as long as two hundred years earlier. Olson gives the "feel of the land," and shares with the reader the "sense of fulfill-ment" that comes to men traveling in the bush together." The book is also valuable for the information it contains on the state of native Indian culture in Canada in the late 1950s.

300   Olson, Sigurd F. *Sigurd F. Olson's Wilderness Days*. New York: Knopf, 1972.
A collection that resounds with echoes of the author's long familiarity with the Quetico-Superior and the Canadian North. These chapters, chosen from earlier books by Olson's friends and readers, range from exciting adventures running rapids or racing storms to contempla-tive musings on what it means to identify with the land. Each vignette of immediate experience in a particular place, clear and vivid in itself, acquires depth and resonance from Sigurd Olson's memories of past encounters and knowledge of Indian legends and tales of the voya-geurs. Whether he writes of being alone or with compan-ions, he evokes the immense silence of the wilderness broken only by the pounding of waves or the cry of a loon. There are occasional elegiac notes as he recalls a now dead companion or describes abandoned homesteads or mining camps, but the country he knows so well remains pristine and youthful, a place where time stands still because here, at least, the continent has not been "tamed and harnessed to the will of man."

301   Proennek, Richard.   *One Man's Wilderness; From the Journals and Photograph Collection of Richard Proennek.*   Ed. Sam Keith.   Anchorage, AK: Alaska Northwest Publishing, 1973.
The fascinating story of a man who quit his job in 1970 at the age of fifty and retired to the Alaskan wilderness. Proennek went into the wild with only simple tools and his own skill to build a cabin that would provide him the shelter he needed to thrive living alone with only the animals of the north for company.   His journals tell of living simply without waste in harmony with his surroundings.   They also reveal the sordidness of those who enter the wilderness only to desecrate its pristine beauty: campers who claim to love the wilderness but leave in their wake beer cans, bottles, and assorted junk; hunters who leave most of the meat from their kills lying on the mountainside or dress it so poorly that it rots in the hanging.   Proennek's journals are written in simple, straightforward prose well suited to convey the integrity of his life.

302   Pyle, Robert Michael.   *Wintergreen: Rambles in a Ravaged Land.*   New York: Scribner's, 1986.
An evocative representation of the Willapa Hills in the coniferous rain forest of Washington State coupled with an examination of the effects of intensive forestry on the land and an indictment of the logging industry. Ecologist Robert Pyle observes the subtle shadings of color and the distinctive patterns of wet weather in this evergreen country and catalogues the behaviors of its extraordinary fauna of snails and salamanders.   His background in forestry enables him to speak with authority as well as outrage about hills "butchered" and "savaged" by clearcutting.   Like other conservationist writers he finds himself engaged in a story of conflict between narrow cut-and-run economic values and broader values that recognize the worth of forest beyond timber; but he is most compelling in his analysis of how the awakening ecological concerns of the citizenry are lulled by deceptive advertising, outright mendacity, and delusive pseudo-science.   In comparing the actual destruction of the land and deliberate extermination of bears and other large mammals by Weyhauser, Crown Zellerbach, Georgia Pacific and others with their manipulative television advertisements, he does much to clarify how environmentally destructive practices can continue unabated in the face of changing public values.

303   Rutstrum, Carl.   *Once Upon a Wilderness.*   New York: Macmillan, 1973.
Informal, personal essays that investigate the benefits of wilderness, the qualities of personality and character necessary for enjoying a wilderness life, and the prospects for wilderness in the future.   From his cabin in the Ontario canoe country veteran outdoorsman Calvin Rutstrum compares past with present and urban with wild to draw some conclusions about what is important in human life. He finds authenticity in the absence of clutter, not only the clutter of things but the clutter of secondhand

feelings and opinions. In exchange for the urban captivi-
ty of "*inescapable, excessive need*" and obsession with
gadgets, time spent in the wilderness offers sensitivity
to the elemental experiences of the sights and sounds of
nature. The wilderness environment encourages "serenity
for gaining knowledge, whether natural or academic." From
the vantage point of wilderness one can be more objective
and philosophical about the news of the day. Wilderness
is not the primitive past, over and done with; it should
be understood as "persevering, ever-pervading nature" that
can and will return whenever and wherever the temporary
incursions of man fall into decay. Rutstrum hopes that
the newer understanding of wilderness as intact ecosystems
will finally end the ravages of multiple-use management
of public lands.

304 Smith, Dwight. *Above Timberline: A Wildlife
Biologist's Rocky Mountain Journal.* New York:
Knopf, 1981.
The "human story of a scientifically trained man who lives
more or less alone in the wilds for a few months, observing and thinking about the environment he is living in."
Dwight Smith takes leave from teaching at Colorado State
University to spend the months between June and November
at a cabin near Brown's Pass on the continental divide in
the Sawatch Range of San Isabel National Forest studying
alpine ecology and filming and taping his experiences for
the New Explorer series. He shares with the reader in
colloquial language not only his ecological observations
but his emotions and life history. Though somewhat
embarrassed by what he fears other scientists would term
anthropomorphism, he yet feels welcomed by the playful,
wordless communication of the small animals--marmots,
martens and chipmunks--that surround his wilderness home.
He muses on the importance of knowledge and understanding
to the sense of belonging in the wilderness, something
sadly lacking in the motorcyclists who ignorantly churn
the fragile alpine soil. Wilderness becomes part of his
religious faith. One of the most moving incidents in the
book is a mass held in his cabin by a visiting priest to
celebrate the creation with a reading of the 104th psalm.

305 Snyder, Gary. *The Practice of the Wild.* San Francisco,
CA: North Point Press, 1990.
Graceful essays on wilderness as place and practice
suffused with ecological courtesy and gratitude. For poet
Gary Snyder wilderness, as a "*place* where wild potential
is fully expressed" in a diversity of "flourishing living
and nonliving beings," is the true home of humankind.
Cultures arise from and depend on the native plants and
animals of a bioregion. Only our centralized industrial
culture imagines itself outside and independent of wild
nature. We can reaffirm our belonging to the assembly of
the wild whole. By practicing awareness of and respect
for the nonhuman members of our local communities, we come
at last to a point where the "blue mountains walk out to
put another coin in the parking meter, and go down to the
7-Eleven." Most of the essays in this book grew out of
Snyder's talks, conversations and workshops over more than

a decade.   Several reflect his knowledge of Buddhism.
Others relate his experiences among native peoples in
Alaska.   Among the most evocative is his retelling and
interpretation of the Eskimo story, "The Woman Who married
a Bear."

306   Stegner, Page. *Outposts of Eden: A Curmudgeon at Large
        in the American West.*   San Francisco, CA: Sierra
        Club Books, 1989.
Accounts of travels through the BLM desert lands and down
the scenic rivers of the American West infused with
satiric humor and the clarity of open spaces.   Page
Stegner introduces this collection of essays by contrasting the claustrophobia of a nightmarish trip he once took
fleeing New York City with his relief on opening his eyes
some forty hours of driving later somewhere in Oklahoma
to "blessed nothing...trackless, pathless space."   He is
a master of creating paradox, irony and humor from the
juxtaposition of opposites. The exquisite adaptations of
desert wildlife are contrasted with the heedlessness of
off-road bikers churning up the burrows of kangaroo rats
and kit foxes as they casually destroy the delicate desert
soil.   He teases mercilessly a group of college students
on a wilderness field trip whose self-righteous environmental opinions are accompanied by a strange lack of
appreciation for their magnificent surroundings.   But he
saves his most angry polemics for the BLM and the Forest
Service, those "federal stewards employed to assist in the
multiple abuse of our public lands."   Page Stegner writes
with passionate love of place in the acerbic tradition of
his father Wallace Stegner, Bernard De Voto, and Edward
Abbey.

307   Sullivan, William L. *Listening for Coyote: A Walk Across
        Oregon's Wilderness.*   New York: Holt, 1988.
A daily journal of a wilderness backpacking adventure
walking thirteen hundred miles from Oregon's westernmost
point at Cape Blanco to its easternmost at Hat Point on
the edge of Hell's Canyon.   William Sullivan walks the
Klamaths, the Cascades, the Ochocos and the Blue Mountains
to learn what wilderness is.   On the way he meets some
wilderness characters--a sentinel for Earth First!, a
Kalmiopsis marijuana grower, and an unconventional forest
ranger with an earring in his ear doing his best to keep
roadless areas roadless.   He also discovers threats to
wilderness everywhere--loud deer hunters in shiny,
oversized ORVs, logging roads slashed through old-growth
forests, and grazing cattle destroying native plants even
in designated wilderness areas. The spirit of wilderness
is more elusive. Partly, it involves solitude: "wilderness is a human word for human absence."   It entails
"daring to get 20 per cent cold and 15 per cent wet" and
facing the possibility of death. Sullivan comes closest
to capturing its mystery in the teasing, invisible
presence of coyotes throughout his trip. They watch and
howl and leave scat messages on the trails, reminding him
that wilderness is the American Indian demigod "Coyote's
magic planet" and that it has been part of him all along.

308   Teale, Edwin Way. *Autumn Across America.* New York: Dodd,
        Mead, 1956.

309   ---. *Journey into Summer.* New York: Dodd, Mead, 1960.

310   ---. *North with the Spring.* New York: Dodd, Mead, 1951.

311   ---. *Wandering Through Winter.* New York: Dodd, Mead,
        1965.
        Four books that make up a naturalist's loving catalogue
        of American wild and rural environments and the plants and
        animals that inhabit them.   Over a period of almost
        fifteen years the author and his wife traveled America to
        record changes in the landscape and wildlife through the
        seasons of the year.   Teale is a noticer.   His smooth,
        short sentences detail what he sees and hears as he
        observes the natural world and listens to the tales of
        people--scientists, park rangers and ordinary folk--he
        meets on the way.   The desire to be inside other forms of
        life and see with their eyes and minds runs like a thread
        through his adventures.

312   Terres, John K.   *From Laurel Hill to Siler's Bog: The
        Walking Adventures of a Naturalist.*   New York:
        Knopf, 1969.
        Sensitive studies of wildlife on the 1,000 acre old Mason
        Farm set aside as a wilderness refuge by the University
        of North Carolina at Chapel Hill.   John Terres was
        encouraged to roam at will over this land while he was
        working on an encyclopedia of North American birds at the
        University.   For seven and one-half years he devotes much
        of his time night and day to "searching the hidden wild
        things of its acres."   To satisfy his naturalist's
        curiosity he endures long hours crouching or lying in
        fields, thickets, canvas blinds or tree platforms in all
        kinds of weather, listening and looking and honing his
        senses razor-sharp.   He is rewarded with some remarkable
        discoveries and develops an intuitive empathy with the
        wild and free.   He live traps a rare golden mouse; he
        observes the courtship display of an almost legendary
        black wild turkey gobbler.   Once he is granted a moving
        glimpse of nature's subtle complexity when he discovers
        that one of a mated pair of red-tailed hawks has placed,
        outside of breeding season, a "fresh sprig of green pine"
        in their abandoned nest, a "badge of ownership to their
        home in the oak."   Terres' book ranks high as a scientific
        and literary tribute to the "healing solitude of the
        still-wild lands."

313   Trimble, Stephen, ed.   *Words from the Land: Encounters
        with Natural History Writing.*   Salt Lake City, UT:
        Gibbs Smith, 1988.
        An anthology of essays by contemporary natural history
        writers who "articulate our neglected connection with the
        rest of the living world in language both passionate and
        thoughtful."   Before putting this book together editor
        Stephen Trimble visited and interviewed each of the
        authors asking about the qualities and purposes of a
        nature writer.   What emerges is the picture of people who

care about the written word, respond personally to the
landscape, have a taste for solitude, believe that nature
counts and are willing to risk conventional truths by
being open to experience.   Trimble selects evocative
essays from most of our best known writers, Edward Abbey,
Barry Lopez and Anne Zwinger, for example.   He also
introduces the reader to some less known contributors to
the genre--Sue Hubbell, David Quammen and John Madson.
Among the most memorable of the pieces included in this
collection is an excerpt from Wendell Berry's little known
1971 work *The Unforseen Wilderness: An Essay on Kentucky's
Red River Gorge*. Berry epitomizes the mysterious belonging of the wilderness experience in the image of a man
standing in his living room in complaisant certainty who
looks up suddenly to see the roof is gone and he inhabits
a "darkness reaching out to the remote lights of the sky."

314   Walker, Theodore J.   *Red Salmon, Brown Bear: The Story
       of an Alaskan Lake*.   New York: World, 1971.
The daily journal of six months alone on a wilderness
island off the coast of Southeastern Alaska.   Biologist
T. J. Walker was commissioned by the New Explorers program
of the American Museum of Natural History to film and
record on audiotape the ecological interactions of plant
and animal life around Lake Eva on Baranof Island, the
glacial drainage system of a coniferous rain forest, for
a television special.   That the book reproduces in print
Walker's oral tapes gives his narrative an exciting
immediacy and an easy familiarity.   For all his expertise
Walker presents himself humorously as a kind of everyman,
steeling himself for the ordeal of emerging from his
sleeping bag on a cold morning, fighting a usually losing
battle with a recalcitrant electric generator, fretting
over a burned thumb and indulging in idiosyncratic
neatnesses in his solitary life. His emotional reactions
to the callous depredation of wildlife and habitat by
humans are sharp and fresh.   He gets sick to his stomach
and enraged at greedy, careless hunters; and he finds our
ignorant anthropocentrism a "little embarrassing," when
we are so obviously as ecologically "derivative and
dependent as the other organisms." By the time he leaves
his by then beloved island, he is convinced that no
wilderness will be protected unless individual people
become personally attentive to nature, "even if it
involves only an insect crawling across your lawn." There
is "more to life than writing and talking."

315   Wallace, David Rains.   *The Dark Range: A Naturalist's
       Night Notebook*.   San Francisco, CA: Sierra Club
       Books, 1978.
An imaginative guide to the nighttime activities of
wildlife in the Yolla Bolly Mountains of California based
on the author's nocturnal forays into the night wilderness
over a period of five years.   David Rains Wallace repeatedly camped and backpacked without special equipment in
the different life zones of the Yolla Bollys, keeping
himself awake for much of many nights, in order to create
a composite picture of a world most humans ignore.   Like
Sally Carrighar he transforms scientific knowledge and

personal observation into narrative through imaginative
empathy with nonhuman life even on its simplest level.
Few writers can so vividly convey what it must feel like
to learn and hunt by smell or evoke the confusion of a
noctuid moth escaping the idle clash of a bear's jaws.
This beautifully illustrated book serves as a primer for
"connecting human sympathies to our nonhuman relatives"
and for learning to listen to the wilderness.

316   Wallace, David Rains.   *The Klamath Knot: Explorations
         of Myth and Evolution.*   San Francisco, CA: Sierra
         Club Books, 1983.
Speculations on the mythological significance of evolution
inspired by a stay in the Klamath wilderness. Evolution
is grounded in science, but false or incomplete interpretations of its meaning have allowed it to be used to
bolster myths that have no basis in fact or reason.   In
relating his imaginative experiences and scientific
explanations of the rocks, water, forests and meadows of
the Klamaths, Wallace uncovers the confused thinking that
lies behind the caricatures that guide popular notions of
evolution as teleological, hierarchal and ruthlessly
competitive. Symbiosis, neoteny and preadaptation are as
characteristic of evolution as competition; and, unlike
earlier myths, the story of evolution is open-ended.
There is no predetermined future, and man is not the end
of the evolutionary process.   The value of wilderness is
that, away from the tamed human environment, it is
possible to catch a glimpse of the mysterious lives of
other beings--both animal and plant--from their own points
of view and in the context of vast evolutionary time. The
wilderness lures us "not to devour us but to remind us
where we are, on a living planet." Wallace's book is a
remarkable imaginative antidote to rampant anthropocentrism.

317   Wallace, David Rains.   *The Untamed Garden and Other
         Personal Essays.*   Columbus, OH: Ohio State
         University Press, 1986.
Observant occasional essays covering a wide range of
wilderness and wildlife topics from the miniature inverte-
brate serengetis of a manure pile in the author's suburban
backyard to the history of wetlands destruction and
conservation in the United States since Colonial times.
Among essays focused on special wild places like the
Okefenokee Swamp or particular species, such as, ravens
or beavers, Wallace includes meditations on nature
broadcasting and nature writing. He explains the recency
of nature writing as a literary genre, not as a historical
accident, but as the corollary of its revolutionary
content with nature as subject rather than setting.
Before Linnaeus nature was generally seen as a "backdrop
to a historical cosmos, or a veneer over a religious one."
All modern nature writing, including television narrative,
has in common appreciative aesthetic responses to the
scientific view of nature.   The natural history writer
translates information into feeling and vision.   This is
certainly true of Wallace himself, whose clear prose,
subtle humor, and imaginative sympathy bring to life the

little known wildernesses that are still "holding up the biosphere."

318   Zwinger, Anne.   *Beyond the Aspen Grove*.   New York: Random House, 1970.
A record of the plants, animals and soils that form the miniature ecosystem of the author's retreat in the mountains of Colorado.   When Anne Zwinger and her husband first buy their forty wilderness acres, they think of it as property they own.   Later, as they live in harmony with their surroundings, they come to see themselves as belonging instead to the land, earning their place in the community only as it becomes something they experience. The book is divided into chapters detailing the interrelationships which make up the nearby micro-environments--the lake, the mountain stream, the meadow, the aspen grove, and the pine forest. A short final chapter centers on the author's favorite spot for contemplation--the lake rock from which it is possible to view the inexorable successive stages of the local ecosystem.   This is a book with a fine sense of place.

319   Zwinger, Anne.   *A Desert Country Near the Sea*.   New York: Harper & Row, 1983.
Records of the author's explorations of the sierras, countryside, towns, and Pacific and gulf coasts of Baja, California.   This is country that Zwinger and her family have visited and loved for many years.   She writes authoritatively about ecosystems as diverse as the mountain meadows of La Laguna or the beach of the Sea of Cortez.   Incidents as simple as blundering into a spider's web as she cuts across country serve to focus a wealth of scientific data and historical information from early explorers and missionaries who wrote with wonder of a little known land.   Many of the plants and animals she encounters are also rendered in her fine drawings and included in the appended list of plants of the Cape region.   Such techniques combine to create a dynamic and multi-layered picture of a place where wild country is still a part of everyday life.

320   Zwinger, Anne.   *Run, River, Run: A Naturalist's Journey Down One of the Great Rivers of the West*.   New York: Harper & Row, 1975.
The story of the author's many trips through Green River country told as a single journey from the river's source in the Wind River Mountains of Wyoming to its confluence with the Colorado in Utah.   This technique enables Zwinger to convey a sense of what she calls "riverness"--the combination of direction and change that make a river seem a living thing.   Unlike thrillseekers who come for the rapids and chaff at stretches of quiet water, she is equally at home gazing at the cold turquoise of a congealed mountain lake, watching in the bow of a canoe for the pillows that betray hidden boulders, or sinking gratefully after a hot day into the opaque silted waters near the river's end.   Only in the section of the river that runs channeled, strung with barbed wire, and banked with dead and bloated cattle through ranch country and

again in the dead waters of Fontenelle Reservoir does her enthusiasm wane. But these places, however desolate, are not wilderness; and it is the healthy land communities of the wild river that inspire the accuracy and grace of her prose and drawings.

321 Zwinger, Anne. *Wind in the Rock*. New York: Harper & Row, 1978.
Distillations of the author's backpacking and horseback trips through the canyon country of southeastern Utah. This is empty, hard, unyielding country where canyon walls of sandstone, limestone or shale drop vertically or are carved into steep overhangs, rattlesnakes are a ubiquitous presence, and water is scarce and difficult to locate. Paradoxically, the toughness that requires the walker to watch her step and to accept responsibility for her own safety inspires a sense of freedom in Zwinger. She leaves the solitude and silence of Grand Gulch, having metaphorically shucked off the carapace that protects her from the communications overload of modern life, with a feeling of renewal and increased sensitivity to the world around her. In a final section she faces directly the question of the value of wilderness and the threats to its continued existence. Wilderness is a "safety valve for civilization," where we can cast aside the materialistic agenda of society to be in harmony with the ancient rhythms of existence. Yet, even at the mouth of Slickrock Canyon, evidence of abuse offends the eye and the nose. Zwinger and her companions must spend an hour cleaning up a filthy campsite before they can camp for the night.

322 Zwinger, Anne, and Edwin Way Teale. *A Conscious Stillness: Two Naturalists on Thoreau's River*. New York: Harper & Row, 1982.
The experiences and conversations of two accomplished nature writers canoeing the Assabet and Sudbury Rivers from their sources in remote marshlands to their confluence in historic Concord. There is an elegiac grace to this book and much of biography and autobiography, for Edwin Way Teale died in 1980 before the text was ready for publication. The final version consists of alternating passages of his preliminary notes and records with Anne Zwinger's comments about the same scenes and incidents. They write of the natural beauty of these New England rivers, of their history, and of the pollution, dams, and highways now on the route that Thoreau travelled more than a century earlier. They also write of themselves, of each other, and of the what it means to be a writer of natural history. Teale ponders their special relationship to nature and their desire to catch and hold their experience in prose; Zwinger remembers with affection their time together and Teale's wide knowledge and humanity. Both authors have the ability to convey to the reader the sights and sounds of the patches of wilderness that still survive among the populated areas of New England.

323 Zwinger, Anne, and Beatrice W. Willard. *Land Above the Trees: A Guide to American Alpine Tundra*. New York: Harper & Row, 1972.

A collaborative effort that translates ecologist Willard's research and scientific knowledge into Zwinger's vital prose and delicate drawings. In preparation for writing this book the two authors studied several alpine areas in the contiguous United States, including Rocky Mountain National Park and places as distant as the White Mountains of New Hampshire and the Cascades of Oregon. A first section describes the ecosystems and inhabitants of a generalized alpine environment; in the second, each chapter is devoted to a particular place. The accurate scientific descriptions of ecological interrelationships and the lovingly detailed attention to specific species convey the visual simplicity and clarity of an illuminated manuscript world. In spite of its inhospitableness to human biology, the authors experience the alpine tundra as an "escape of freedom" for the human spirit. In a brief concluding section they discuss the need to protect the tundra against sheep grazing and the heavy traffic of hiking boots if it is to continue its role in storing and conserving water.

324   Zwinger, Anne Haymond. *The Mysterious Lands: A Naturalist Explores the Four Great Deserts of the Southwest.* New York: Dutton, 1989.
Overview descriptions of the varying scenery and plant and animal life of the Chihuahuan, Sonoran, Mojave and Great Basin Deserts. Anne Zwinger uses a combination of historical references, conversations with desert botanists, biologists and ecologists, and her own sensitivity to the natural world to convey the uniqueness of each of these desert environments. Although many of her desert trips are in the company of others, she is most successful in evoking her experience in the desert solitude. Hidden behind a blind at a water tank in the Cabeza Prieta National Wildlife Refuge, she discovers precision and serenity in bighorn sheep and shares an afternoon nap against her duffel bag with a trusting black-tailed gnatcatcher. Eventually she comes to prefer the desert with its "absences and the big empties" to stifling easy landscapes, "neat-trimmed lawns" and "ruffled curtains of trees." Because of its broad, overall coverage Zwinger's book could serve as background reading for the more specific and localized studies of other desert writers.

AUTHOR INDEX

This index includes entries for contributors of relevant writings to proceedings,
collections, anthologies and special
issues of journals as well as entries for
authors of monographs or separately cited
essays and articles. Contributors are
indexed whether or not they are specifically mentioned in the annotations. The
numbers following authors' names refer to
items in the bibliography.

Abbey, Edward 208, 209,
    236, 237, 238, 239, 242,
    243, 287, 313
Adams, Ansel 116
Ahlgren, Clifford 23
Ahlgren, Isabel 23
Alcock, John 240, 241
Allen, Barry 112
Allen, Durward L. 123
Allen, Robert P. 17
Allen, Thomas B. 24
Allin, Craig W. 25
Anderson, Clinton P. 118,
    122
Andy, J. Ralph 119
Armstrong, Edward A. 17
Armstrong-Buck, Susan 155
Attfield, Robin 156, 170

Babbitt, Bruce 16
Baer, Richard 112
Baker, Ron 124
Bailey, Alfred M. 17
Banks, James T. 131
Bannerman, David 17
Barnett, Philip S. 131
Baron, Robert 241

Bates, Marston 81
Bean, Michael J. 26
Beard, Daniel B. 115, 122
Bergson, Frank 243
Berry, Thomas 99
Berry, Wendall 287, 313
Beston, Henry 242, 244, 287
Birch, Charles 157
Birch, Thomas H. 158
Blackwelder, Eliot 153
Bodi, F. Loraine 131
Bonner, James 118
Bookchin, Murray 107
Borland, Hal 127
Bowen, James 108
Bradley, David 153
Bradley, Nina Leopold 16
Brandborg, Stewart M. 121
Braun, Ernest 145
Bratton, Susan P. 230
Brennan, Andrew 159
Brewer, Jo 127
Broome, Harvey 29
Brooks, Paul 1, 27, 28, 118,
    119, 122, 127, 245
Brower, David 2, 21, 115,
    116, 117, 118, 119, 122,

169, 242
Brown, Leslie H. 17
Brown, Michael 90
Brown, Tom, Jr. 107
Buchheister, Carl W. 119
Burke, Albert E. 121
Burke, C. John 265
Burke, James A. 118
Butler, Lisa 230

Cahalane, Victor H. 17
Cahn, Robert 30
Cain, Stanley L. 115, 122
Callicott, J. Baird 3, 93,
    160, 161, 162, 163, 164,
    165, 166, 167, 170
Carlson, Allen 168
Carr, Archie 127
Carr, James A. 118
Carrighar, Sally 242, 246,
    247
Carson, Rachel 82, 242, 243,
    248, 287
Carter, Everett 116, 122
Carter, T. Donald 17
Carver, John A., Jr. 118
Catton, William R., Jr. 107
Chase, Alston 31
Chisholm, Alec 17
Clary, David 32
Cliffe, Edward 118, 121
Cobb, John B., Jr. 96, 157
Cohen, Michael J. 109, 110,
    111
Cohen, Michael P. 33
Collier, John 117
Commission on Americans
    Outdoors 219
Condliffe, John B. 118, 122
The Continental Group 220
Cook, Sam 249
Cottam, Clarence 17
Cowan, Robert McTaggert 115
Cowles, Raymond B. 115
Crafts, Edward C. 118
Crandell, Harry 231
Crandell, Rick 230

Daly, Herman E. 107
Daniel, John 210
Darling F. (Frank) Fraser 4,
    17, 83, 115, 122
Darlington, David 5
Dasmann, Raymond 16, 84
Davis, Bruce 22
Davis, Eric 107
Day, Lincoln H. 118
Dearden, Philip 22
Delacour, Jean 17

Demars, Stanford E. 34
Devall, Bill 125, 169
De Voto, Bernard 133
Dickey, James 107
Dillard, Annie 242, 250,
    265, 287, 313
Disilvestro, Roger L. 35
Dodd, Elizabeth M. 131
Dolin, Eric Jay 222
Douglas, William O. 106,
    116, 122, 134, 135, 242,
    251, 252, 253
Dubos, Rene 205
Dumbrell, John 22
Durr, Robert 137

Eckleberry, Donald 17
Ehrenfeld, David 85, 95, 96,
    100
Ehrlich, Anne 86
Ehrlich, Gretel 254, 265,
    313
Ehrlich, Paul 86, 96, 120
Eidsvik, J. 22
Eiseley, Loren 211, 212,
    213, 242, 255
Eliot, Wayne 90
Elliot, Robert 170
Ervin, Keith 36
Evans, Brock 120
Evans, Howard Ensign 265
Everson, William 107

Faulkner, William 214
Fayad, Elizabeth A. 131
Finch, Robert 256, 257, 265,
    313
Fletcher, Colin 145
Ford, Laurence D. 80
Fowles, John 265
Fox, Stephen 7, 37
Fox, Warwick 171
Fradkin, Philip L. 38, 39
Frankena, W. K. 172
Franklin, Jerry F. 231
Freeman, Orville L. 121
French, Roderick 112
Fritzell, Peter 3
Frome, Michael 40, 121, 136,
    258
Fuller, R. Buckminster 120
Furtman, Michael 259
Futrell, J. William 131

Gabrielson, Ira N. 17
Gare, Aaron 170
Gauvin, Aime 127
Gibson, William 118
Gilbert, Bil 260, 261

Gilbert, Rudolph W. 121
Gilham, G. E. 118
Gillard, E. Thomas 17
Gilliam, Harold 118
Gilligan, James P. 118, 119, 121
Gladden, James N. 41
Glover, James A. 8
Godfrey-Smith, William 173
Goodin, Robert 170
Goodpaster, Kenneth E. 174
Gould, Samuel 119
Graber, Linda H. 42
Graf, William L. 43
Graham, Frank, Jr. 44, 127
Grapard, Alan G. 107
Graves, John 262
Gray, Brian F. 131
Gray, Elizabeth Dodson 175
Green, Harry W. 80
Gruchow, Paul 263, 264

Haag, John 45
Haagen-Smit, Arie J. 118
Haas, Glen 223, 224, 231
Haines, John 101, 137, 215, 265, 287
Halpern, David 265
Hammitt, William E. 225
Hammond, John L. 176
Haneman, W. Michael 96
Hardin, Garrett 106, 107
Hargrove, Eugene 95, 177, 178
Hartzog, George B., Jr. 118, 121
Harvey, D. Michael 131
Hay, John 107, 242, 266, 267, 268, 287, 313
Hays, Samuel P. 46
Henberg, Martin 138
Hendee, John C. 90, 126, 130
Herman, Eric 224
Higbee, Edward 116
Hill, Thomas E., Jr. 179
Hines, Lawrence 121
Hoagland, Edward 242, 265, 269, 270, 287, 313
Hochbaum, H. Albert 17, 120
Hocker, Philip 131
Hooker, C. A. 170
Hoover, Helen 271, 272
Hope, Jack 127
Hubbell, Sue 273, 313
Hunt, Charles B. 112
Huth, Hans 47, 116

Jackson, Henry M. 120
Jeffers, Robinson 169, 216, 217
Johnson, Huey D. 16
Josephy, Alvin M., Jr. 127
Junkin, Elizabeth Darby 242

Kafka, Gregory S. 170
Kaplan, Rachel 226
Kaplan, Stephen 226, 227
Kaufman, Sharon 16
Keiter, Robert B. 131
Kellert, Stephen R. 228, 229
Kellow, Aynsley 22
Kerr, Clark 118
Kesselheim, Alan S. 274
Kheel, Marti 180
Kilgore, Bruce M. 117, 118, 122
King, Thomas F. 131
Kinnell, Galway 107
Kloefkorn, William 137
Knopf, Alfred 153
Krutch, Joseph Wood 87, 116, 122, 127, 139, 242, 275, 276, 277, 278, 287

La Bastille, Anne 9, 279, 280
La Chapelle, Dolores 107
Lack, David 17
Laycock, George 127
Leake, Chauncey 118, 119
Ledig, F. Thomas 80
Leduc, Thomas 137
Lehmann, Scott 181
Lehmberg, Paul 281
Leopold, A. Starker 117, 122
Leopold, Aldo 88, 140, 141, 242, 243, 282, 283
Leopold, Carl 16
Leopold, Estella B. 16, 121
Leopold, Frederic 16
Leopold, Luna B. 16, 115, 118, 122
LePawsky, Albert 118, 122
Leydet, Francois 119
Limerick, Patricia Nelson 48
Line, Les 127
Lister, Robert 153
Littlebird, Larry 96
Livingston, John A. 102
Lockhart, William J. 131
Lopez, Barry (Holstun) 103, 265, 285, 287, 313
Losos, Jonathan B. 80
Lovelock, J. (James) E. 89, 96
Lucas, Robert C. 22, 230, 231
Lueders, Edward 286

Lyon, Thomas J. 287
Lyons, Oren R. 90

Macinko, George 120
Madson, John 127, 288, 313
Magraw, Daniel Barstow 131
Manes, Christopher 49
Manning, Robert C. 22
Margolis, John D. 10
Marshall, Robert 142, 143,
  289
Marston, Otis 153
Martin, Russell 50
Martin, Vance 90
Marx, Wesley 127
Matthews, Freya 182
Matthiessen, Peter 51, 287,
  313
McCloskey, Maxine 120, 121
McCloskey, Michael 52, 53
McClure, Michael E. 96
McConnell, Grant 54, 119,
  122
McDonald, Corry 55
McGinley, Patrick 131
McKibben, Bill 104
McNamee, Thomas 230
McNulty, Faith 127
McPhee, John 11, 242, 243,
  290, 313
Meine, Curt 3, 12
Meinertzhagen, Richard 17
Merchant, Carolyn 56, 57,
  183
Metcalf, Lee 119
Midgley, Mary 170, 205
Milbraith, Lester W. 232
Millar, Constance J. 80
Miller, Peter 184
Milton, John 265
Milton, John P. 120, 291
Mitchell, John G. 13, 128
Montagu, Ashley 118, 122
Moran, Emilio F. 107
Moss, Frank L. 121
Murie, Adolph 292
Murie, Margaret 137, 242,
  293, 294
Murie, Olaus 17, 144, 153,
  294, 295
Myers, Norman 91, 96, 107

Nabham, Gary 313
Naess, Arne 107, 185, 186
Narvesan, Jan 170
Nash, Roderick (Fraser) 3,
  58, 59, 60, 61, 90, 107,
  112, 120, 121, 145, 146,
  147, 223, 231, 242

Nelson, J. Gordon 22
Nelson, Richard K. 265, 296
Nelson, Urban C. 120
Norgaard, Richard B. 96
Norris, Kenneth S. 80, 121
Norse, Elliot A. 92
Norton, Bryan G. 93, 95, 96,
  187, 188, 189
Ntombela, Magqubu 90

Oandasen, William 107
Odell, Rice 62
Oelschlaeger, Max 63
Ogburn, Charlton, Jr. 242
Olson, Sigurd F. 116, 118,
  122, 127, 148, 242, 297,
  298, 299, 300
O'Riordan, Jon 22
Osborn, Fairfield 119
Ottinger, Richard 1. 120
Outdoor Recreation Resources
  Review Commission 221
Owings, Margaret Wentworth
  118

Palmer, Tim 64, 65, 66
Parker, Walter 137
Partridge, Ernest 190
Passmore, John 191
Pearl, Mary C. 95
Peattie, Donald Culcross 287
Penfold, Joseph W. 116, 153
Perrin, Noel 265
Peterson, Roger Tory 17
Pettingill, Olin Sewall, Jr.
  17
Piel, Gerald 116, 122
Pister, Edwin P. 3
Player, Ian 90
Price, Monroe 137
Proennek, Richard 301
Pyle, Robert Michael 302

Quammen, David 313

Randall, Aln 93, 96
Rapp, Dennis A. 121
Rasmassen, Boyd L. 121
Rausch, Robert 115
Reed, Peter 192
Regan, Donald H. 93
Regan, Tom 193
Reynolds, T. Eric 118
Ribbens, Dennis 3
Rienow, Robert 121
Roberts, David 243
Roberts, Rebecca S. 230
Rodman, John 194, 265
Rolston, Holmes, III 3, 95,

170, 195, 196, 197, 198
Rossman, Betty B. 233
Routley, Richard 170, 199
Routley, Val 199
Rozin, Skip 127
Runte, Alfred 67, 68
Rutstrum, Calvin 303

Sadley, Barry 22
Sagoff, Mark 113
Sax, Joseph L. 129, 131
Saylor, John p. 116
Schaefer, Paul 149
Schoenfeld, Clarence A. 130
Schoenwald-Cox, Christine M.
  90
Schrepfer, Susan R. 69
Schreyer, Richard 231, 235
Schwarz, William 122
Scott, Douglas 90
Scott, Neil 234
Scoyen, Eivind 116, 122
Searle, R. Newell 70
Sears, Paul 116, 118
Sessions, George 107, 125,
  169
Sewell, W. R. Derrick 22
Shepard, Paul 107
Shrader-Frechette, Kristin
  200
Simon, David 131
Siri, William 118
Siry, Joseph 112
Smith, Dwight 304
Smith, Spencer M., Jr. 119,
  122
Sober, Eliot 93
Soleri, Paolo 107
Soper, J. Dewey 17
Soucie, Gary 127
Soulé, Michael 94, 95, 96
Snyder, Gary 105, 169, 218,
  305
Spilhaus, Athelstan 119
Spurr, Stephen H. 119, 122
Squillance, Mark 131
Stankey, George H. 22, 231,
  235
Stegner, Page 306
Stegner, Wallace 3, 14, 117,
  122, 150, 151, 152, 153,
  243
Stoddard, Charles H. 118
Stoddard, Herbert L., Sr. 17
Stone, Christopher, D. 106,
  201, 202
Strong, Douglas H. 15
Sullivan, William L. 307
Sumner, Lowell 117

Talbot, Lee Merriman 117,
  120
Tallmadge, John 3
Tanner, Thomas 16
Taylor, Paul 203, 205
Taylor, Walter P. 17
Teale, Edwin Way 127, 243,
  308, 309, 310, 311, 322
Terres, John K. 17, 312
Terrie, Phil 112
Tobias, Michael 107
Tober, James A. 71
Thompson, Janna L. 170, 204
Trevelyan, G. M. 72, 115,
  122
Trimble, Stephen 313
Turner, Frederick 73

Udall, Stewart 74, 75, 116,
  119, 122
Ulehla, Z. Joseph 233

Van De Veer, Donald 205
Vest, Jay Hansford C. 206
Vickery, Jim Dale 18

Walker, Theodore J. 314
Wallace, David Rains 315,
  316, 317
Walsh, Richard 224
Warren, Karen J. 207
Warren, Mary Anne 170
Warren, Virginia 170
Watkins, Charles, Jr. 76
Watkins, T. H. 19, 20, 76
Wayburn, Peggy 118
Wayburn, Edgar 118
Weedon, Richard B. 120
Western, David 95
White, Lynn, Jr. 77
Willard, Beatrice W. 323
Wilson, Edward O. 95, 96,
  97, 98, 265
Worster, Donald 78
Wurster, Catharine Bauer 116

Yambert, Paul A. 114
Young, Robert A. 230
Young, Stanley P. 17

Zahniser, Howard 116, 117,
  122, 154
Zaslowsky, Dyan 79
Zeldin, Melvin 127
Zwinger, Anne (Haymond) 242,
  265, 313, 318, 319, 320,
  321, 322, 323, 324

TOPIC INDEX

Abbey, Edward 48
Adirondack State Park (NY)
  8, 149, 279, 280
aesthetic values 47, 72,
  142, 143, 165, 168, 177,
  221, 226, 227, 317
African Americans 222
Alaska 4, 8, 25, 30, 101,
  120, 137, 148, 215, 236,
  245, 253, 258, 285, 291,
  292, 293, 295, 296, 301,
  305, 314, 317
Alaska Coalition 30
Alaska National Interest
  Lands 30, 79
anthropocentric values 91,
  93, 96, 98, 179, 187, 188,
  189, 191
Appalachia 136
Arches National Park 237
Arctic 284, 295
Arizona 145, 208, 209,
  238, 240, 241, 260, 275,
  276, 278, 283, 324
Assabet River (NH) 322
Athabasca River (Canada) 274
Atlantic Coast 244, 248,
  256, 257, 266, 267, 268
Audubon Expedition Institute
  109

Baja California (Mexico)
  127, 245, 277, 319
Big Bend National Park (TX)
  251
Big Horn Mts.(MT, WY) 264
biocentric values 126, 172,
  174, 193, 203
biodiversity 84, 85, 86, 92,

93, 94, 96, 98, 189
Boundary Waters Canoe Area
  Wilderness (MN) *See also*
  Quetico-Superior. 23, 44,
  70, 127, 245, 259, 297,
  298
Brazos River (TX) 262
Brooks Range (AK) 253, 258,
  289, 291, 295
Broome, Harvey 252, 258
Brower, David 2, 11, 15, 75

California 2, 5, 65, 66, 68,
  216, 217, 315, 316, 317
California condor 5, 71
Canada *See also Quetico-
  Superior*. 261, 284, 295,
  299 303
Canyonlands National Park
  (UT) 245
canyon country 237, 320
Cape Cod (MA) 244, 256, 257,
  266, 267, 268
Carson, Rachel 1, 15, 75
Cascade Mts. (OR, WA) 253,
  307, 323
Churchill River (Canada) 299
Christianity 73, 77, 99,
  156, 191
clearcutting 136, 302
Colorado 304, 318, 324
Colorado River 38, 50
Commoner, Barry 15
Concord River (NH) 322
conservation biology 80, 94,
  95
conservation organizations
  71
conservationists 7, 9, 15,

17, 28, 37, 42, 43, 44, 54
cost-benefit analysis 173,
    198, 229
cultural change 45, 46, 57,
    59, 61, 63, 99, 105, 232

dams 38, 50, 64, 66, 145
Darling, F. (Frank) Fraser
    4, 17
Darling, Jay (Ding) 24
Death Valley (CA) 239
deep ecology 49, 105, 107,
    125, 169, 171, 182, 185,
    186, 199
deserts 48, 121, 236, 237,
    324
De Voto, Bernard 14
Dinosaur National Monument
    (CO, UT) 39, 64, 153
Douglas, William O. 258, 293

ecocentric values See also
    deep ecology. 108, 155,
    163, 166, 173, 185, 194,
    195, 196, 197, 216, 217,
    278, 305
ecofeminism 175, 180, 207
ecology 6, 78, 81, 84, 85,
    88, 120, 159, 161, 166,
    248, 282, 283
Ecosophy T 185
ecosystem preservation and
    restoration 22, 90, 92,
    94, 95
ecotage 49, 208, 209
educational values 147, 154,
    188, 189
emotion, role in valuing
    nature 180, 192, 203
endangered species 35, 51,
    91, 94
environmental ethics 59, 62,
    63, 95, 167, 178, 183
ethical character 179, 190,
    203
ethical pluralism 159, 184,
    201, 202
Everglades National Park
    (FL) 28, 252
evolution 87, 89, 97, 211,
    212, 213, 227, 240, 241,
    255, 278, 316
extinction 51, 86, 96

fact/value dichotomy 161,
    164, 178, 196
Florida 31, 127
Forest Service, U. S. 32,
    40, 136

forests and forestry 23, 32,
    36, 136
freedom and wilderness 49,
    158, 204, 236, 237, 239,
    305

Gaia hypothesis 89, 109
Georgia 317
Gila Wilderness (NM) 282,
    283
Glacier National Park (MT)
    131, 239
Glacier Peak Wilderness (WA)
    28, 253
Glen Canyon (UT) 50, 64,
    152, 237
global warming 104
Grand Canyon (AZ) 145, 236,
    285
Grand Gulch Canyon (UT) 321
grazing on the public lands
    39, 43, 76, 133
Great Smoky Mountains
    National Park (NC,TN) 136,
    245, 252
Green River (CO,UT,WY) 64,
    153, 238, 239, 320

Hell's Canyon (OR) 307
heritage values 116, 121,
    138, 152, 176, 214, 299
Hudson Bay (Canada) 295
humanism 100
hunting 35, 123, 124, 128,
    249, 260, 282, 283, 296

Ickes, Harold L. 15, 19
Illinois 282, 283, 288
Iowa 283, 288

Jeffers, Robinson 63
John's Canyon (UT) 321
Judaism 156, 191

Kansas 288
kinship with nonhumans 87,
    97, 139, 163, 211, 212,
    213, 260, 296, 305, 315
Kalmiopsis Mts. (OR) 307
Klamath Mts. (CA,OR) 307,
    316
Krutch, Joseph Wood 10, 48

land ethic 151, 161, 162,
    200, 282, 283
law and legislation 25, 26,
    106, 131
Leopold, A. Starker 4
Leopold, Aldo 3, 6, 12, 15,

16, 63, 160, 164, 165

MacKaye, Benton 7
Macmillan, Eben 5
Macmillan, Ian 5
Maine 127, 252, 266, 270
Marshall, Robert 7, 8, 13,
   29, 149
Massachusetts 244, 256, 257,
   266, 267, 268
Michigan 128
Middle West 263, 288
Minnesota 23, 41, 127, 245,
   259, 264, 269, 271, 272,
   297, 298, 300
Missouri 128, 273, 283, 288
Mohave Desert (CA) 324
monkeywrenching 49, 208, 209
Montana 128, 264
motorized recreation 41,
   127, 129
Mount Hood (OR) 323
Mount Katahdin (ME) 252
Mount McKinley National Park
   (AK) 245, 292
Mount Ranier (WA) 323
Mount Washington (NH) 323
Murie, Margaret 9, 293, 294
Murie, Olaus 7, 17, 252,
   293, 294

national forests 36, 79, 136
national parks 67, 79, 129,
   131
National Park Service 34,
   131, 136
National Resource Lands of
   the BLM 76, 79
National Wilderness
   Preservation System 43, 79
National Wildlife Federation
   24
native peoples 73, 162, 295,
   296, 299, 305
Nebraska 264, 288
New England 57, 270, 322
New Hampshire 269, 322, 323
New Melones Dam (CA) 66
New Mexico 55, 282, 283
North Carolina 136, 245, 312
North Cascades National Park
   28
North Dakota 288
Northwest Territories
   (Canada) 261, 274
nature writing 264, 287, 317

Oberholtzer, Ernest 70
Ochocos Mts. (OR) 307

Okefenokee Swamp (GA) 317
Oklawaha River (FL) 28
old-growth forests 36, 92,
   302
Olson, Sigurd F. 18, 70,
   252, 258
Olympic National Park (WA)
   245, 253, 323
Oregon 36, 92, 253, 307, 316
Ozark Mts. 261, 273
Outdoor Challenge Program
   226

Pacific coast 216, 217, 285,
   307
Pacific Northwest 36, 92,
   302
Pennsylvania 260, 261
pesticides 82
population growth 83, 85,
   90, 95, 104, 119
process philosophy 155, 157
psychological values 132,
   142, 221, 225, 226, 233,
   285

Quetico-Superior (MN,
   Ontario) 18, 70, 148, 245,
   249, 271, 272, 281, 297,
   298, 300
quantum theory 163

radical environmentalism 49
recreation 138, 219, 224
Redwood National Park (CA)
   69
religion 37, 45, 61, 73,
   99, 156, 157, 175
rights for nonhumans 61,
   106, 181
rivers 64, 238
Rocky Mts. (CO) 304, 318,
   323
Rutstrum, Calvin 18

Salmon River (ID) 253
San Juan River (UT) 238, 321
Save-the-Redwoods League 69
scientific values 91, 115
Sequoia National Park (CA)
   69, 246
Sierra Club 2, 33, 69
Sierra Nevada (CA) 65, 253,
   323
Snyder, Gary 63
solitude 206, 225, 262
Sonoran Desert (AZ) 240,
   241, 264, 275, 276, 278,
   324

spiritual values 99, 197,
    250, 278
Stanislaus River (CA) 66
Steer Gulch Canyon (UT) 321
Stegner, Wallace 20
stewardship 156, 191
Superior-Quetico Council 70

tallgrass prairie 263, 264,
    288
Tellico Dam 136
Tennessee 245
Teton Mts. (WY) 247, 294
Texas 127, 128, 251, 262,
    269

Uinta Mts. (UT, WY) 39
Utah 39, 50, 208, 209, 306,
    320, 321, 324

values classifications 52,
    146, 147, 173, 183, 198,
    223, 231
Vermont 270
Virgin Islands 245
Virginia 250

Washington 28, 36, 92, 253,
    302
The West 4, 38, 39, 43, 133,
    150, 152, 239, 306
Western world picture 56,
    57, 63, 73, 77, 99, 100,
    170, 175, 191
wetlands 317
Whirlwind Gulch (UT) 321
White Mts. (NH) 323
Whitehead, Alfred North 155,
    157
Wilderness Act of 1964 21,
    25, 40, 52, 53, 55, 119
wilderness as a human right
    134, 135
wilderness management 22,
    31, 130, 141, 230, 231
Wilderness Society 29, 132,
    140
wildlife 26, 35, 51, 102,
    228, 246, 247, 269, 271,
    272, 284, 292, 312
wildlife management 71, 95,
    123, 124, 126, 130
Willapa Hills (WA) 302
Wind River Mts. (WY) 252,
    320
Wisconsin 4, 283, 286
wolves 103
women 9, 56, 273, 279, 280
Wyoming 254, 294, 320

Yellowknife River (Yukon)
    261
Yellowstone National Park
    (ID, MT, WY) 31, 258
Yolla Bolly Mts. (CA) 315
Yosemite National Park (CA)
    34, 68, 239
Yukon (Canada) 261

Zahniser, Howard 7, 21, 75,
    149

About the Compiler

JOAN S. ELBERS is a Reference Librarian and Associate Professor at Montgomery College in Maryland. Her publications include a bibliography and articles on topics in history and literature.